Dedication

Then Jesus told them, **"A prophet is honored everywhere except in his own hometown and among his relatives and his own family**." Mark 6:4 New Living Translation

This dedication goes out to Stan's family, friends, and supporters worldwide. This document you are about to read has been paid for in Stan's own blood and the sweat and tears of those that believed and supported Stan in all his struggles to help the American people break free of our dependence on fossil fuels. If Stan would have had his way the world would be so much different than it is today and for many good reasons.

Special thanks goes out to all those worldwide that have honored Stan as a teacher sent from God Himself to lead us into a better way of providing ourselves and families with the energy we have so desperately learned to depend on.

If you are like me, you have printed this document out a number of times and read it over and over again trying to solve the mystery Stan left us all.

This printed book will save you the trouble of printing it out and make Stan's work more permanent, more widely read, researched and respectable, I hope.

This printing is from public domain and all proceeds go towards non-profit "Free Energy" organizations.

Special thanks to Brother Andrew that has helped keep this PDF file alive and well on his Stan Meyer's dedication website: http://waterpoweredcar.com/stanmeyer.html

Captain Willy

Director Lablanc Inventors Academy

Preface

Reprinted from the article: The car that ran on water
Written by Dean Narciso
The Columbus Dispatch • Sunday July 8, 2007

Stanley Meyer's bizarre death at age 57 ended work that, if proved valid, could have ended reliance on fossil fuels.

People who knew him say his work drew worldwide attention: mysterious visitors from overseas, government spying and lucrative buyout offers.

His death sparked a three-month investigation that consumed and fascinated Grove City police.

"Meyer's death was laced with all sorts of stories of conspiracy, cloak-and-dagger stories," said Grove City Police Lt. Steve Robinette, lead detective on the case.

If Stephen Meyer was shocked at his twin brother's collapse and death, he was equally amazed at the Belgians' response the next day.

"I told them that Stan had died and they never said a word," he recalled, "absolutely nothing, no condolences, no questions. "I never, ever had a trust of those two men ever again."

Today, Stanley Meyer is featured on numerous Internet sites. A significant portion of the 1995 documentary It Runs on Water, narrated by science-fiction writer Arthur C. Clarke and aired on the BBC, focuses on his "water fuel cell" invention.

James Robey wants a permanent place for Meyer in his Kentucky Water Fuel Museum.

"He was ignored, called a fraud and died without his small hometown even remembering him with so much as a plaque," Robey wrote in his self-published book Water Car.

Meyer had euphoric highs and humiliating defeats. He was kind and generous yet paranoid and suspicious. He would be hailed as a visionary and a genius. He also would be sued and declared a fraud.

As many of his more than 20 patents expire this year, and gasoline prices hover around $3 per gallon, there is growing interest in his inventions. But it remains unclear how much was true science and how much was science fiction. See the whole article at:

http://www.dispatch.com/content/stories/local/2007/07/08/hydroman.ART_ART_07-08-07_A1_4V77MOK.html

See http://waterpoweredcar.com/stanmeyer.html for more videos, the free PDF for this book and much more.

And visit: http://fuel-efficient-vehicles.org/energy-news/?page_id=925

It Runs On Water Video: https://www.youtube.com/watch?v=2zX0P3E5c_c

The Birth of New Technology

Water Fuel Cell
Technical Brief

Explaining the Hydrogen Fracturing Process on how to use water as a new fuel-source

" Meets All Energy Needs "

Copyright: Public Domain
ISBN 978-1-329-50241-3

Published by Lablanc Inventors Academy
A Non-Profit Green Energy Charity
www.lablancinventors.org

Written by Stanley Allen Meyer

Stanley A. Meyer, Inventor
3792 Broadway
Grove City, Ohio 43123
1-614-871-4173
Fax: 1-614-871-8075

WATER FUEL CELL

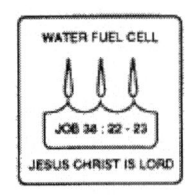

The Birth of New Technology

Patents Granted To Date   Stanley A. Meyer

4,389,981	Hydrogen gas injector system for internal combustion engine (U.S.A.)
4,613,779	Electrical pulse generator (U.S.A.)
4,421,474	Hydrogen gas burner (U.S.A.)
1,231,872	Hydrogen injector system (CDA)
1,233,379	Hydrogen gas injector for internal combustion engine (CDA)
1,228,833	Gas electrical hydrogen generator (CDA)
1,227,094	Hydrogen/air & non-combustible gas mixing combustionsystem (CDA)
4,613,304	Gas electrical hydrogen generator (USA)
1,235,669	Controlled hydrogen gas flame (CDA)
4,275,950	Light-guide lens (USA)
1,234,774	Hydrogen generator system (USA)
3,970,070	Solar heating system (USA)
1,234,773	Resonant cavity hydrogen generator that operates with a pulse voltage electrical potential (CDA)
4,265,224	Multi-stage solar storage system (USA)
1,213,671	Electrical particle generator (CDA)
4,465,455	Start-up / shut-down for a hydrogen gas burner (USA)
4,798,661	Gas generator voltage control circuit (USA)
4,826,581	Controlled process for the productionof thermal energy from gases and apparatus useful therefore (Hydrogen Fracturin Process) (PCT)
5,149,407	Process and apparatus for the production of fuel gas and the enhanced release of thermal energy from such gas (Electronic interfacing for the Hydrogen Fracturing Process) (Resonant Action) (USA) (WFC Project 423 DA)
0101761	Controlled hydrogen gas flame (EPO)
1577992	Controlled hydrogen Gas flame (JPO)
0086439	Hydrogen gas injector system for internal combustion engine (EPO)
1584224	Hydrogen Injection System (JPO)
4,936,961	Method For the production of a Fuel Gas "Electrical Polarization Process" (U.S.A.)
1,694,782	Resonant Cavity For Hydrogen Generator (JPO)
5,293,857	Hydrogen gas fuel and management system for an internal combustion engine utilizing hydrogen gas fuel (U.S.A.)

Other U.S. & Foreign Patents Pending

Refer to WFC Profit Sharing Certificate Prospectus when considering purchasing a WFC Dealership or obtaining a WFC Profit Sharing Certificate

About the Author

Stanley A. Meyer

Stanley A. Meyer, a businessman and free-lance inventor, lives in Grove City, Ohio. His scientific and engineering background covers many fields of endeavors: Heart Monitors for the medical profession, the Validator System System for the banking institution, the Nivax and Actar System for the oceanography field, and the "EBED" concept for Star Wars, to mention a few. And, now, Mr. Meyer has developed the Water Fuel Cell technology to help solve the energy crisis. Many energy patents have been granted to him over the years.

Stanley A. Meyer founded and served as chairman of several high technology business and co-sponsored other business activities in the international market place.

While continuing to set up Water Fuel Cell business entity and inventing, Stanley A. Meyer has begun working on a book entitled "With the Lord, There is Purpose" describing his "faith-walk" with the Lord to fulfill end-time prophecy. He continues his speaking engagements throughout the world.

Recipient Awards of Merit:
1990 - Who's Who of American Inventors
1991 - 1992 Who's Who Of Entrepreneurs U.S.A.
1992 - Who's Who of American Inventors
1993- Who's Who of American Inventors of the Year Award
1994- Who's Who of American Inventors

Publications of Authorship
Raum & zeit: U.S.: Vol. 2 No. 1, 1990; Vol 3 No. 4, 1992
Raum & zeit: Europe: 9 Jahrgang Nr 44; 9 Jahrgang Nr 48; 9 Jahrgang Nr 50
Explore: U.S.: Vol 3 No. 4, 1992; Vol 4 No. 2, 1993

Speaker of Request:
1989 SAFE International Congress for Free Energy, Einsiedeln, Switzerland
1990 International Extraordinary Science, Colorado Springs, Colorado
1991 International Global Clean Energy Congress, Geneva, Switzerland
1991 International Clobal Science Congress, Daytona Beach, Florida
1993 International Symposium on New Energy, Denver, Colorado
1994 International Solar Expo 94, Ukiah, California

WATER FUEL CELL

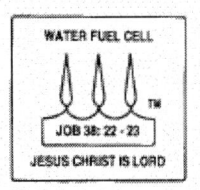

The Birth of New Technology

WFC Tech-Brief

Table of Contents

Book History	Page Locator
Scientific Paragon	Preface
Sec. 1) Memo 420: Hydrogen Fracturing Process Date of Entry: 01/25/90	25
Sec. 2) Memo 421: Quenching Circuit Technology Date of Entry: 01/25/90	11
Sec. 3) Memo 422DA: WFC Hydrogen Gas Management System Date of Entry: 04/15/91	50
Sec. 4) Memo 423DA: Water Fuel Injection System Date of Entry: 07/03/91	13
Sec. 5) Memo 424: Atomic Energy Balance of Water Date of Entry: 11/14/91	13
Sec. 6) Memo 425: Taper Resonant Cavity Date of Entry: 08/13/92	07
Sec. 7) Memo 426: VIC Matrix Circuit Date of Entry: 07/07/93	24
Sec. 8) Memo 427: Voltage Wave-Guide Date of Entry: 08/10/93	15
Sec. 9) Memo 428: Exhaust Air Reclaimer Date of Entry: 06/18/94	08
Sec. 10) Memo 429: Optical Thermal Lens Date of Entry: 11/03/95	13
Sec 11) Memo 430: Steam Resonator Date of Entry: 5/18/96	13
Appendix A: Table of Tabulation .... Appx A	04
Appendix B: Glossary of Application Notes .... Appx B	01

Memo WFC 420

WATER FUEL CELL

Hydrogen Fracturing Process ... using Water as Fuel.

Over the Years man has used water in many ways to make his life on Earth more productive. Why not,now, use water as fuel to power our cars, heat our homes, fly our planes or propel spaceships beyond our galaxy? Biblical prophesy foretells this event.

After all, the energy contained in a gallon of water exceeds 2.5 million barrels of oil when equated in terms of atomic energy. Water, of course, is free, abundant, and energy recyclable.

The Hydrogen Fracturing Process dissociates the water molecule by way of voltage stimulation, ionizes the combustible gases by electron ejection and, then, prevents the formation of the water molecule during thermal gas ignition...releasing thermal explosive energy beyond "normal" gas burning levels under control state ... and the atomic energy process is environmentally safe.

The Hydrogen Fracturing Process is systematically activated and performed in the following way:

Section 1

RE: Hydrogen Fracturing Process   Memo WFC 420

Hydrogen Fracturing Process

Method

Using "Voltage Potential" to stimulate the water molecule to produce atomic energy on demand

Operational Parameters

Pulsing Transformer

The pulsing transformer (A/G) steps up the voltage amplitude or voltage potential during pulsing operations. The primary coil is electrically isolated (no electrical connection between primary and secondary coil) to form Voltage Intensifier Circuit (AA) Figure (1-1). Voltage amplitude or voltage potential is increased when secondary coil (A) is wrapped with more turns of wire. Isolated electrical ground (J) prevents electron flow from input circuit ground.

Blocking Diode

Blocking Diode (B) prevents electrical "shorting" to secondary coil (A) during pulse-off time since the diode "only" conducts electrical energy in the direction of the schematic arrow.

LC Circuit

Resonant Charging Choke (C) in series with Excitor-array (E1/E2) forms an inductor-capacitor circuit (LC) since the Excitor-Array (ER) acts or performs as an capacitor during pulsing operations, as illustrated in Figure (1-2) as to Figure (1-1).

The Dielectric Properties (insulator to the flow of amps) of natural water (dielectric constant being 78.54 @ 25c) between the electrical plates (E1/E2) forms the capacitor (ER). Water now becomes part of the Voltage Intensifier Circuit in the form of "resistance" between electrical ground and pulse-frequency positive-potential...helping to prevent electron flow within the pulsing circuit (AA) of Figure 1-1.

Stanley A. Meyer

RE: Hydrogen Fracturing Process Memo WFC 420

The Inductor (C) takes on or becomes an Modulator Inductor which steps up an oscillation of an given charging frequency with the effective capacitance of an pulse-forming network in order to charge the voltage zones (E1/E2) to an higher potential beyond applied voltage input.

The Inductance (C) and Capacitance (ER) properties of the LC circuit is therefore "tuned" to resonance at a certain frequency. The Resonant Frequency can be raised or lowered by changing the inductance and/or the capacitance values. The established resonant frequency is, of course, independent of voltage amplitude, as illustrated in Figure (1-3) as to Figure (1-4).

The value of the Inductor (C), the value of the capacitor (ER), and the pulse-frequency of the voltage being applied across the LC circuit determines the impedance of the LC circuit.

The impedance of an inductor and a capacitor in series, Z series is given by

$$Z\ series = (Xc - Xl\ ) \qquad (Eq\ 1)$$

Where

(Eq 2) \qquad\qquad (Eq 3)

$$Xc = \frac{1}{2\pi fc} \qquad\qquad Xl = 2\pi fl$$

The Resonant Frequency (F) of an LC circuit in series is given by

$$F = \frac{1}{2\pi \sqrt{LC}} \qquad (Eq\ 4)$$

Ohm's Law for LC circuit in series is given by

$$Vt = I\ Z \qquad (Eq\ 5)$$

Stanley A. Meyer

RE: Hydrogen Fracturing Process Memo WFC 420

LC Voltage

The voltage across the inductor (C) or capacitor (ER) is greater than the applied voltage (H). At frequency close to resonance, the voltage across the individual components is higher than the applied voltage (H), and, at resonant frequency, the voltage V_T across both the inductor and the capacitor are theoretically infinite. However, physical constraints of components and circuit interaction prevents the voltage from reaching infinity.

The voltage (VL) across the inductor (C) is given by the equation

(Eq 6)

$$V_l = \frac{V_t \, X_l}{(X_l - X_c)}$$

The voltage (VC) across the capacitor is given by

(Eq 7)

$$V_c = \frac{V_t \, X_c}{(X_l - X_c)}$$

During resonant interaction, the incoming unipolar pulse-train (H) of Figure (1-1) as to Figure (1-5) produces an step-charging voltage-effect across Excitor-Array (ER), as illustrated in Figure (1-3) and Figure (1-4). Voltage intensity increases from zero 'ground-state' to an high positive voltage potential in an progressive function. Once the voltage-pulse is terminated or switched-off, voltage potential returns to "ground-state" or near ground-state to start the voltage deflection process over again.

Voltage intensity or level across Excitor-Array (ER) can exceed 20,000 volts due to circuit (AA) interaction and is directly related to pulse-train (H) variable amplitude input.

RE: Hydrogen Fracturing Process   Memo WFC 420

RLC Circuit

Inductor (C) is made of or composed of resistive wire (R2) to further restrict D.C. current flow beyond inductance reaction (XL), and, is given by

(Eq 8)

$$Z = \sqrt{R_I^2 + (X_I - X_c)^2}$$

Dual-inline RLC Network

Variable inductor-coil (D), similar to inductor (C) connected to opposite polarity voltage zone (E2) further inhibits electron movement or deflection within the Voltage Intensifier Circuit. Movable wiper arm fine "tunes" "Resonant Action" during pulsing operations. Inductor (D) in relationship to inductor (C) electrically balances the opposite voltage electrical potential across voltage zones (E1/E2).

VIC Resistance

Since pickup coil (A) is also composed of or made of resistive wire-coil (R1), then, total circuit resistance is given by

(Eq 9)

$$Z = R_I + Z_2 + Z_3 + R_E$$

Where, R_E is the dielectric constant of natural water.

Ohm's Law as to applied electrical power, which is

(Eq 10)

$$E = IR$$

Where

(Eq 11)

$$P = EI$$

Stanley A. Meyer

RE: Hydrogen Fracturing Process   Memo WFC 420

Whereby

Electrical power (P) is an linear relationship between two variables, voltage (E) and amps (I).

Voltage Dynamic

Potential Energy

Voltage is "electrical pressure" or "electrical force" within an electrical circuit and is known as "voltage potential". The higher the voltage potential, the greater "electrical attraction force" or "electrical repelling force" is applied to the electrical circuit. Voltage potential is an "unaltered" or "unchanged" energy-state when "electron movement" or "electron deflection" is prevented or restricted within the electrical circuit.

Voltage Performs Work

Unlike voltage charges within an electrical circuit sets up an "electrical attraction force; whereas, like electrical charges within the same electrical circuit encourages an "repelling action". In both cases, electrical charge deflection or movement is directly related to applied voltage. These electrical "forces" are known as "voltage fields" and can exhibit either a positive or negative electrical charge.

Likewise, Ions or particles within the electrical circuit having unlike electrical charges are attracted to each other. Ions or particle masses having the same or like electrical charges will move away from one another, as illustrated in Figure (1-6).

Furthermore, electrical charged ions or particles can move toward stationary voltage fields of opposite polarity, and, is given by Newton's second Law

(Eq 12)

$$A = \frac{F}{M}$$

Where

The acceleration (A) of an particle mass (M) acted on by a Net Force (F).

Stanley A. Meyer

RE: Hydrogen Fracturing Process　　　　　　　　　　　　　　　　Memo WFC 420

Whereby

Net Force (F) is the "electrical attraction force" between opposite electrically charged entities, and, is given by Coulomb's Law

(Eq 13)
$$F = \frac{qq'}{R^2}$$

Whereas

Difference of potential between two charges is measured by the work necessary to bring the charges together, and, is given by

(Eq 14)
$$V = \frac{q}{e^R}$$

The potential at a point due to a charge (q) at a distance (R) in a medium whose dielectric constant is (e).

Atomic Interaction to Voltage Stimulation

Atomic structure of an atom exhibits two types of electrical charged mass-entities. Orbital electrons having negative electrical charges (-) and a nucleus composed of protons having positive electrical charges (+). In stable electrical state, the number of negative electrically charged electrons equals the same number of positive electrically charged protons...forming an atom having "no" net electrical charge.

Whenever one or more electrons are "dislodged" from the atom, the atom takes on a net positive electrical charge and is called a positive ion. If an electron combines with a stable or normal atom, the atom has a net negative charge and is called a negative ion.

Voltage potential within an electrical circuit (see Voltage Intensifier Circuit as to Figure 1-1) can cause one or more electrons to be dislodged from the atom due to opposite polarity attraction between unlike charged entities, as shown in Figure (1-8) (see Figure 1-6 again as to Figure 1-9) as to Newtons's and Coulomb's Laws of electrical force (RR).

Stanley A. Meyer　　　　　　　　　　　　　　　　　　　　　　　　　1-6

RE: Hydrogen Fracturing Process   Memo WFC 420

The resultant electrical attraction force (qq') combines or joins unlike atoms together by way of covalent bonding to form molecules of gases, solids, or liquids.

When the unlike oxygen atom combines with two hydrogen atoms to from the water molecule by accepting the hydrogen electrons (aa' of Figure 1-7), the oxygen atoms become "net" negative electrically charged (-) since the restructured oxygen atom now occupies 10 negative electrically charged electrons as to only 8 positive electrically charged protons. The hydrogen atom with only its positive charged proton remaining and unused, now, takes on a "net" positive electrical charge equal to the electrical intensity of the negative charges of the two electrons (aa') being shared by the oxygen atom...satisfying the law of physics that for every action there is an equal and opposite reaction. The sum total of the two positive charged hydrogen atoms (++) equaling the negative charged oxygen atom (- -) forms a "no" net electrical charged molecule of water. Only the unlike atoms of the water molecule exhibits opposite electrical charges.

Voltage Dissociation of The Water Molecule

Placement of a pulse-voltage potential across the Excitor-Array (ER) while inhibiting or preventing electron flow from within the Voltage Intensifier Circuit (AA) causes the water molecule to separate into its component parts by, momentarily, pulling away orbital electrons from the water molecule, as illustrated in Figure (1-9).

The stationary "positive" electrical voltage-field (E1) not only attracts the negative charged oxygen atom but also pulls away negative charged electrons from the water molecule. At the same time , the stationary "negative" electrical voltage field (E2) attracts the positive charged hydrogen atoms. Once the negative electrically charged electrons are dislodged from the water molecule, covalent bonding (sharing electrons) ceases to exist, switching-off or disrupting the electrical attraction force (qq') between the water molecule atoms.

The liberated and moving atoms (having missing electrons) regain or capture the free floating electrons once applied voltage is switched-off during pulsing operations. The liberated and electrically stabilized atom having a net electrical charge of "zero" exit the water bath for hydrogen gas utilization.

Dissociation of the water molecule by way of voltage stimulation is herein called "The Electrical Polarization Process".

Stanley A. Meyer

RE: Hydrogen Fracturing Process   Memo WFC 420

Subjecting or exposing the water molecule to even higher voltage levels causes the liberated atoms to go into a "state" of gas ionization. Each liberated atom taking-on its own "net" electrical charge. The ionized atoms along with free floating negative charged electrons are, now, deflected (pulsing electrical voltage fields of opposite polarity) through the Electrical Polarization Process ...imparting or superimposing a second physical-force (particle-impact) unto the electrically charged water bath. Oscillation (back and forth movement) of electrically charged particles by way of voltage deflection is hereinafter called "Resonant Action", as illustrated in Figure (1-10).

Attenuating and adjusting the "pulse-voltage-amplitude" with respect to the "pulse voltage frequency", now, produces hydrogen gas on demand while restricting amp flow.

Laser Interaction

Light-emitting diodes arranged in a Cluster-Array (see Figure 1-11) provides and emits a narrow band of visible light energy into the voltage stimulated water bath, as illustrated in Figure (1-13) as to Figure (1-12). The absorbed Laser Energy (Electromagnetic Energy) causes many atoms to lose electrons while highly energizing the liberated combustible gas ions prior to and during thermal gas-ignition. Laser or light intensity is linear with respect to the forward current through the LEDS, and, is determined by

$$Rs = \frac{V_{in} - V_{led}}{I_{led}}$$
(Eq 15)

Where

I_{led} is the specified forward current (typically 20ma. per diode); V_{led} is the LED voltage drop (typically 1.7 volts for red emitters).

Ohm's Law for LED circuit in parallel array, and is given by

$$P\,watts = Vcc \; It$$
(Eq 16)

Where

(It) is the forward current through LED cluster-Array: Vcc is volts applied (typically 5 volts).

Stanley A. Meyer

RE: Hydrogen Fracturing Process   Memo WFC 420

Whereby

Laser or light intensity is variable as to duty cycle on/off pulse-frequency from 1Hz to 65 Hz and above is given by

(Eq 17)

$$Le \sqrt{\frac{(ION)^2 \times T1}{T1 + T2}}$$

Le is light intensity in watt ; T1 is current on-time; T2 is current off-time; and (ION)=RMS value of load current during on-period.

Injecting Laser Energy into the Electrical Polarization Process and controlling the intensity of the light-energy causes the Combustible Gases to reach a higher energy-state (electromagnetically priming the combustible gas ions) which, in turn, accelerates gas production while raising gas-flame temperatures beyond "normal" gas-burning levels.

Injecting "Electromagnetically Primed" and "Electrically Charged" combustible gas ions (from water) into other light-activated Resonant Cavities further promotes gas-yield beyond voltage/laser stimulation, as illustrated in Figure (1-16) as to Figure (1-18).

Electron Extraction Process

Exposing the displaced and moving combustible gas atoms (exiting waterbath and passing through Gas Resonant Cavity (T), Figure (1-17) as to Figure (1-18) to another or separate pulsating laser energy-source (V) at higher voltage levels (E3/E4) causes more electrons to be "pulled away" or "dislodged" from the gas atoms, as illustrated in Figure (1-15) as to Figure (1-8).

The absorbed Laser Energy "forces" or "deflects" the electrons away from the gas atom nucleus during voltage-pulse Off-Time. The recurring positive voltage-pulse (k) attracts (qq') the liberated negative electrically charged electrons to positive voltage zone (E3). While, at the same time, the pulsating negative electrical voltage potential (E4) attracts (qq') the positive electrical charged nucleus.

The Positive Electrical Voltage Field (E3) and Negative Electrical Voltage Fields (E4) are

RE: Hydrogen Fracturing Process   Memo WFC 420

triggered "Simultaneously" during the same duty-pulse.

Electron Extraction Circuit (BB) of Figure (1-14) removes, captures, and consumes the "dislodged" electrons (from the gas atoms) to cause the gas atoms to go into and reach "Critical-State", forming highly energized combustible gas atoms having missing electrons. Resistive values (R4, R6, R7, and dialectic constant of gas Rg) and isolated electrical ground (W) prevents "electron-flow" or "electron deflection" from occurring within circuit (BB) during pulsing operations (at resonant frequency) and, therefore, keeps the gas atoms in critical-state by "NOT" allowing electron replacement to occur or take place between the moving gas atoms.

The "dislodged" negative charged electrons are "destroyed" or "consumed" in the form of "heat" when Amp Consuming Devise (S) (such as a light bulb) is positive electrically energized during alternate pulsing operations. Laser activated or laser primed gas ions repels the "dislodged" electrons being consumed, as illustrated in Figure (1-8) as to Figure (1-20). The Electron Extraction Process (BB) is, hereinafter, called "The Hydrogen Gas Gun" and is placed on top of a Resonant Cavity Assembly, as illustrated in Figure (1-17) as to Figure (1-18).

Thermal Explosive Energy

Exposing the expelling "laser-primed" and "electrically charged" combustible gas ions (exiting from Gas Resonant Cavity) to a thermal-spark or heat-zone causes thermal gas-ignition, releasing thermal explosive energy (gtnt) beyond the Gas-Flame Stage, as illustrated in Figure (1-19) as to (1-18).

Thermal Atomic interaction (gtnt) is caused when the combustible gas ions (from water) fail to unite or form a Covalent Link-up or Covalent Bond between the water molecule atoms, as illustrated in Figure (1-19). The oxygen atom having less than four covalent electrons (Electron Extraction Process) is unable to reach "Stable-State" (six to eight covalent electrons required) when the two hydrogen atoms seeks to form the water molecule during thermal gas ignition.

The absorbed Laser energy (Va, Vb and Vc) weakens the "Electrical Bond" between the orbital electrons and the nucleus of the atoms; while, at the same time, electrical attraction-force (qq'), being stronger than "Normal" due to the lack of covalent electrons, "Locks Onto" and "Keeps" the hydrogen electrons. These "abnormal" or "unstable" conditions causes the combustible gas ions to over compensate and breakdown into thermal explosive energy (gtnt). This Atomic Thermal-

Stanley A. Meyer

RE: Hydrogen Fracturing Process   Memo WFC 420

Interaction between highly energized combustible gas ions is hereinafter called "The Hydrogen Fracturing Process."

By simply attenuating or varying voltage amplitude in direct relationship to voltage pulse-rate determines Atomic Power-Yield under controlled state.

Rocket Propulsion

Add-on Resonant Cavities (placed beneath the Hydrogen Gas Gun Assembly) arranged in parallel to vertical Cluster-Array increases the atomic Energy-Yield of the Hydrogen Fracturing Process undergoing thermal gas-ignition, as illustrated in Figure (1-22) as to Figure (1-18). This Cluster-Assembly or Cluster-form is, hereinafter, called "The water powered rocket engine".

Prolonged-rocket-flights carrying heavier payloads is achieved by liquefying the "specially treated" combustible gas ions (laser primed oxygen gas atoms having missing electrons and laser primed hydrogen gas atoms) under pressure in separate fuel tanks affixed to a Rocket Engine, as illustrated in Figure (1-21). Rocket thrust is now controlled by the flow rate of the combustible ionized gases entering the combustion chamber of the rocket engine once gas-ignition occurs.

In Summation

The Hydrogen Fracturing Process simply triggers and releases atomic energy from natural water by allowing highly energized sub-critical combustible gas ions to come together during thermal gas ignition. The Voltage Intensifier Circuit brings on the "Electrical Polarization Process" that switches off the covalent bond of the water molecule without consuming amps. The Electrical Extraction Circuit not only decreases the mass size of the combustible gas atoms; but, also, and at the same time produces "electrical energy" when the liberated electrons are directed away from the Hydrogen Gas Gun Assembly.

The Hydrogen Fracturing Process has the capability of releasing thermal explosive energy up to and beyond 2.5 million barrels of oil per gallon of water under controlled state...which simply prevents the formation of the water molecule during thermal gas ignition...releasing thermal explosive energy beyond the normal gas combustion process. The Hydrogen Fracturing Process is environmentally safe.

RE: Hydrogen Fracturing Process   Memo WFC 420

The Hydrogen Fracturing Process is design-variable to retrofit to any type of energy consuming devise since the Hydrogen Gas Gun can be reduced to the size of an auto spark plug or a gas injector port of a fighter aircraft or enlarged to form a rocket engine. Prototyping determines operational parameters. The Hydrogen Fracturing Process is registered and certified under the Patent Cooperation Treaty Act via foreign grant license #492680 issued July 10, 1989 and foreign grant license #490606 issued Nov. 15, 1988 by the United States of America as to Hydrogen Fracturing Process U.S. patent #4,826,581 issued May 2, 1989, Electrical Polarization Process U.S. Patent #4,936,961 issued June 26, 1990, Resonant Cavity Voltage Intensifier Circuit (VIC) U.S. Patent 5,149,407 issued Sept. 22, 1992, and other U.S. patents pending under the Patent Cooperation Treaty Act (PCT) Worldwide. (see WFC "Patents Granted To Date").

Stanley A. Meyer

RE: WFC Hydrogen Fracturing Process	Memo WFC 420

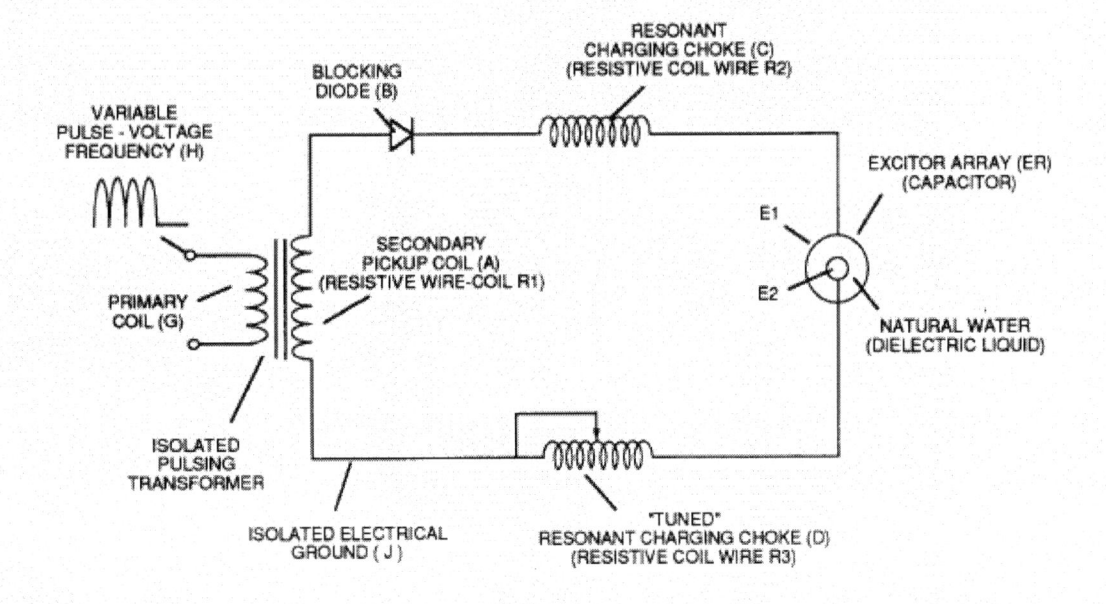

FIGURE 1-1: VOLTAGE INTENSIFIER CIRCUIT (AA)

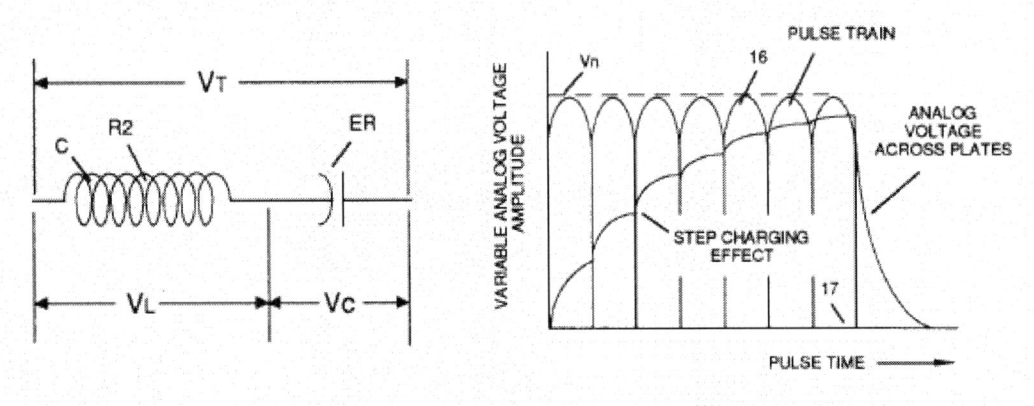

FIGURE 1-2: LC CIRCUIT

FIGURE 1-3: APPLIED VOLTAGE TO PLATES

Stanley A. Meyer

RE: WFC Hydrogen Fracturing Process Memo WFC 420

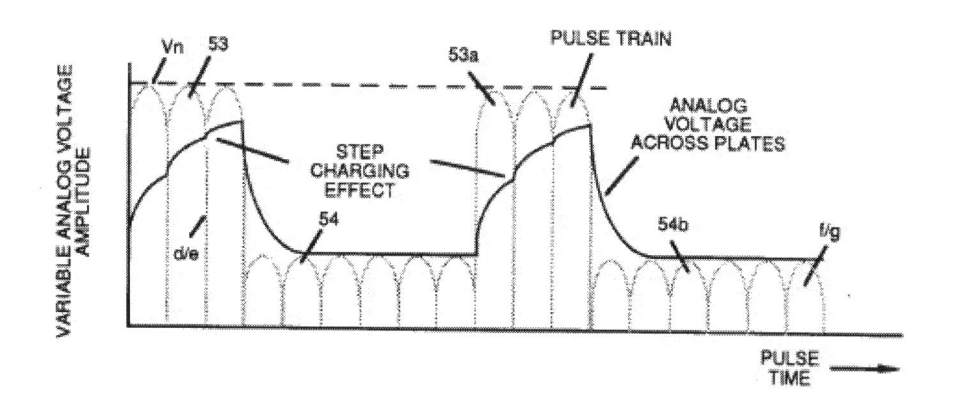

FIGURE 1-4: APPLIED VOLTAGE TO RESONANT CAVITY

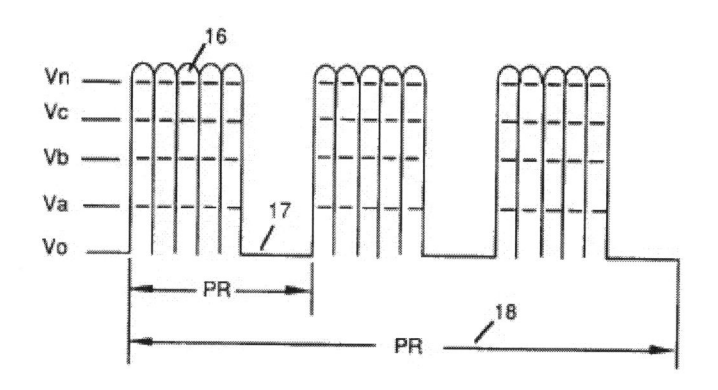

FIGURE 1-5: VARIABLE AMPLITUDE GATED UNIPOLAR PULSE-FREQUENCY DYNAMICALLY CONTROLS HYDROGEN GAS-YIELD ON DEMAND WHILE INHIBITING AMP FLOW

RE: WFC Hydrogen Fracturing Process Memo WFC 420

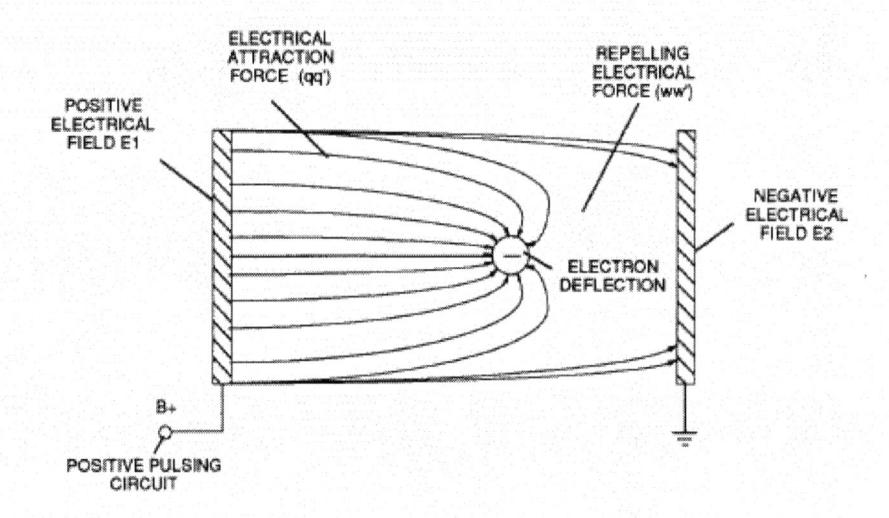

FIGURE 1-6: VOLTAGE POTENTIAL PERFORMING WORK

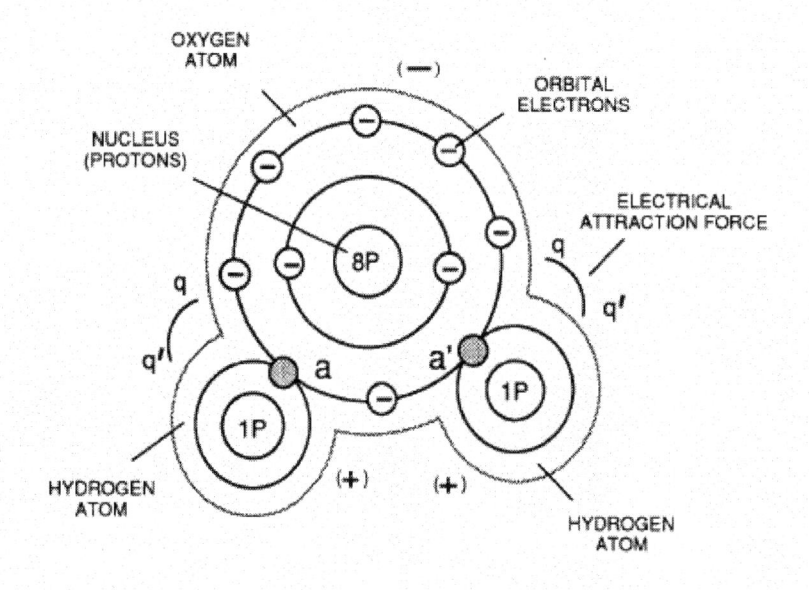

FIGURE 1-7: ELECTRICAL CHARGES OF THE WATER MOLECULE

Stanley A. Meyer 1-15

RE: WFC Hydrogen Fracturing Process   Memo WFC 420

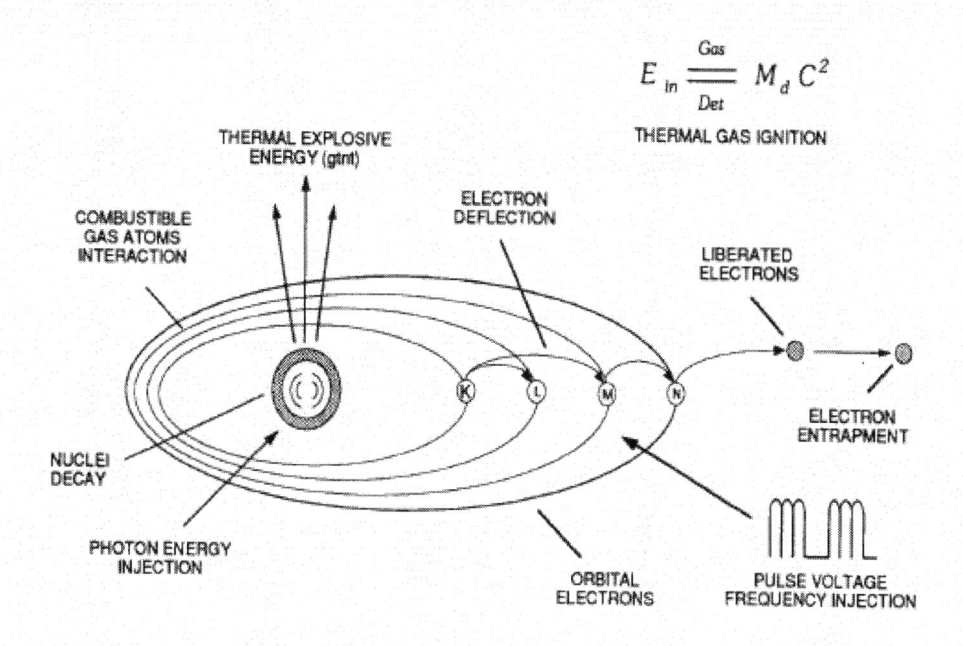

$$E_{in} \underset{Det}{\overset{Gas}{=\!=\!=}} M_d C^2$$

THERMAL GAS IGNITION

FIGURE 1-8: HYDROGEN FRACTURING PROCESS

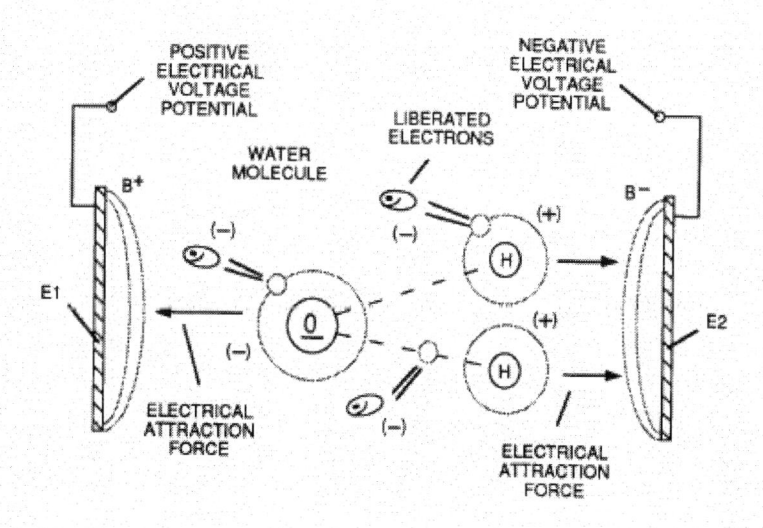

FIGURE 1-9: ELECTRICAL POLARIZATION PROCESS

Stanley A. Meyer   1-16

RE: WFC Hydrogen Fracturing Process   Memo WFC 420

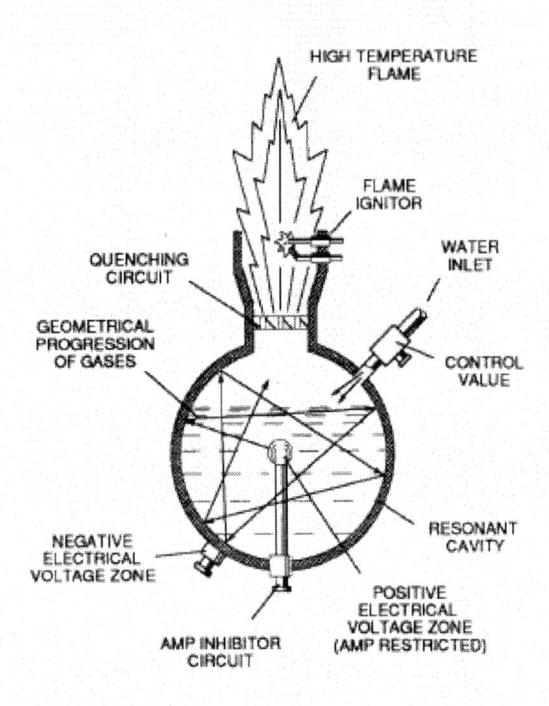

FIGURE 1-10: ELECTRICAL VOLTAGE ZONES
FORMING RESONANT CAVITY

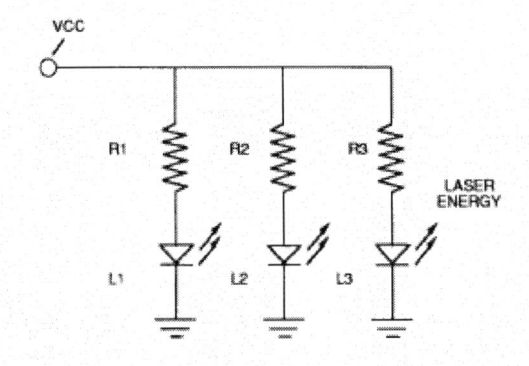

FIGURE 1-11: LED CLUSTER ARRAY

Stanley A. Meyer   1-17

RE: WFC Hydrogen Fracturing Process memo WFC 420

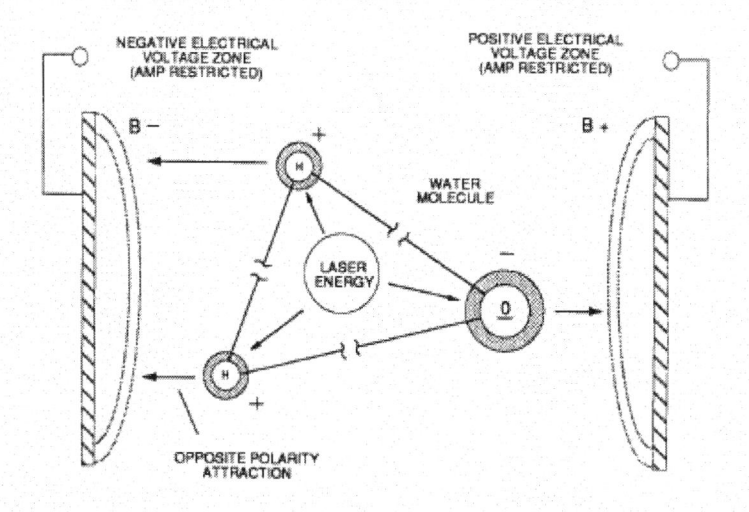

FIGURE 1-12: PHOTON ENERGY AIDS RESONANT ACTION

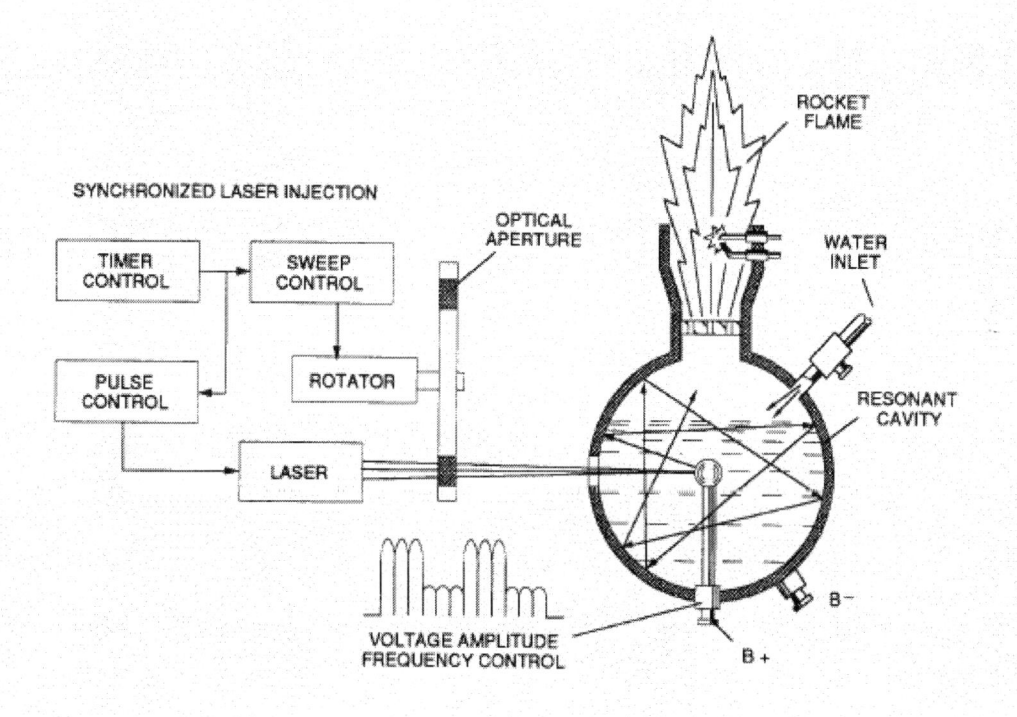

FIGURE 1-13: LASER INJECTED RESONANT CAVITY

RE: WFC Hydrogen Fracturing Process   Memo WFC 420

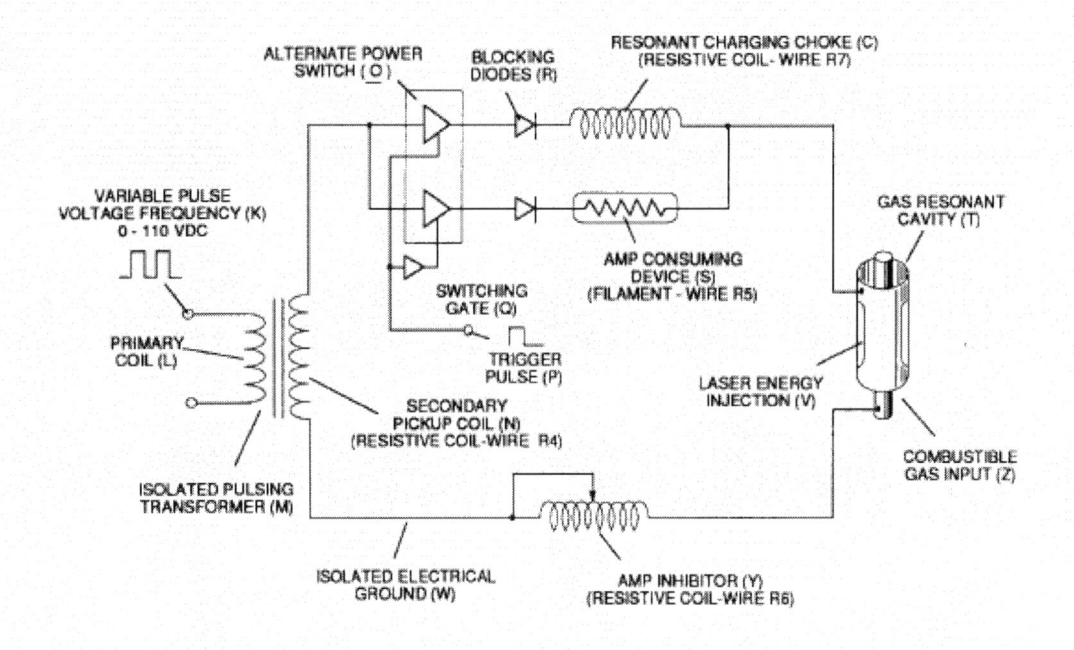

FIGURE 1-14: ELECTRON EXTRACTION CIRCUIT (BB)

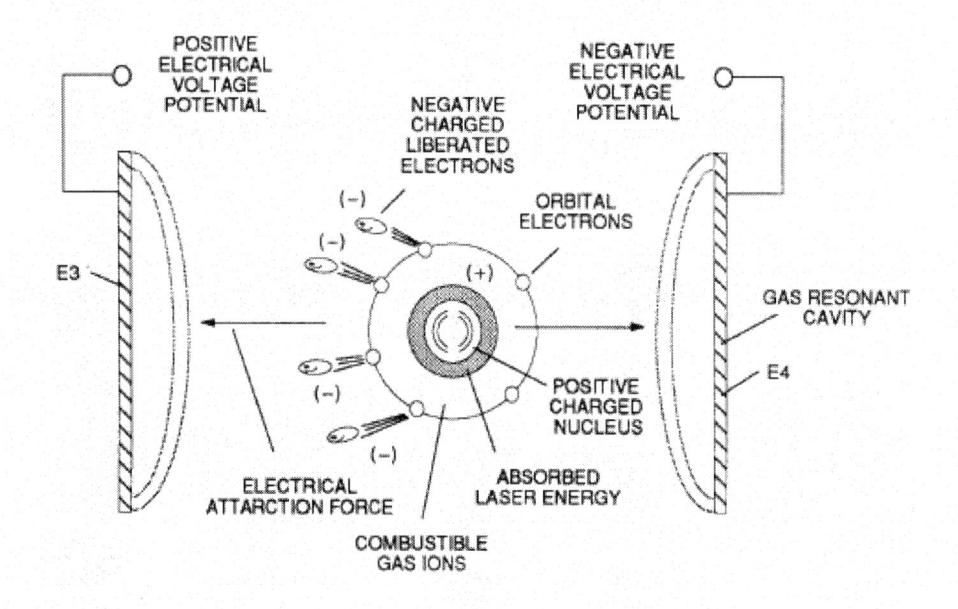

FIGURE 1-15: DESTABILIZING COMBUSTIBLE GAS IONS

Stanley A. Meyer   1-19

RE: WFC Hydrogen Fracturing Process   Memo WFC 420

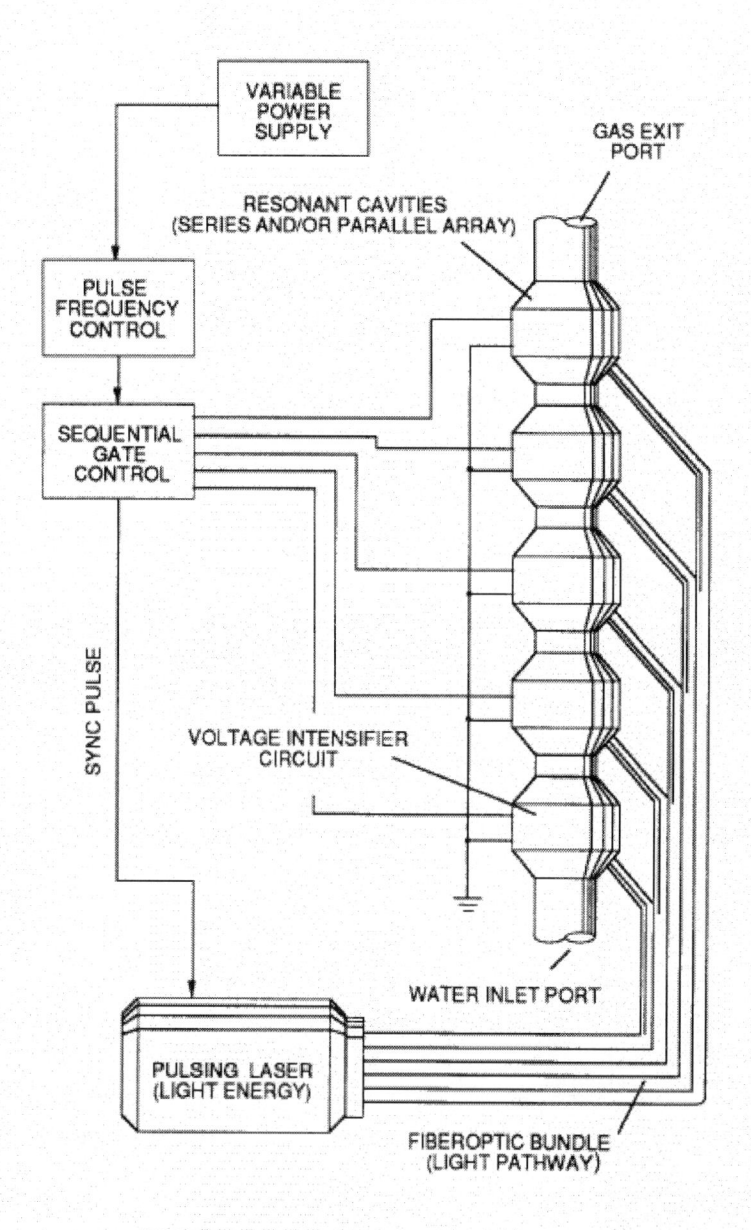

FIGURE 1-16: POWER LOAD DISTRIBUTION

Stanley A. Meyer   1 - 20

RE: WFC Hydrogen Fracturing Process Memo WFC 420

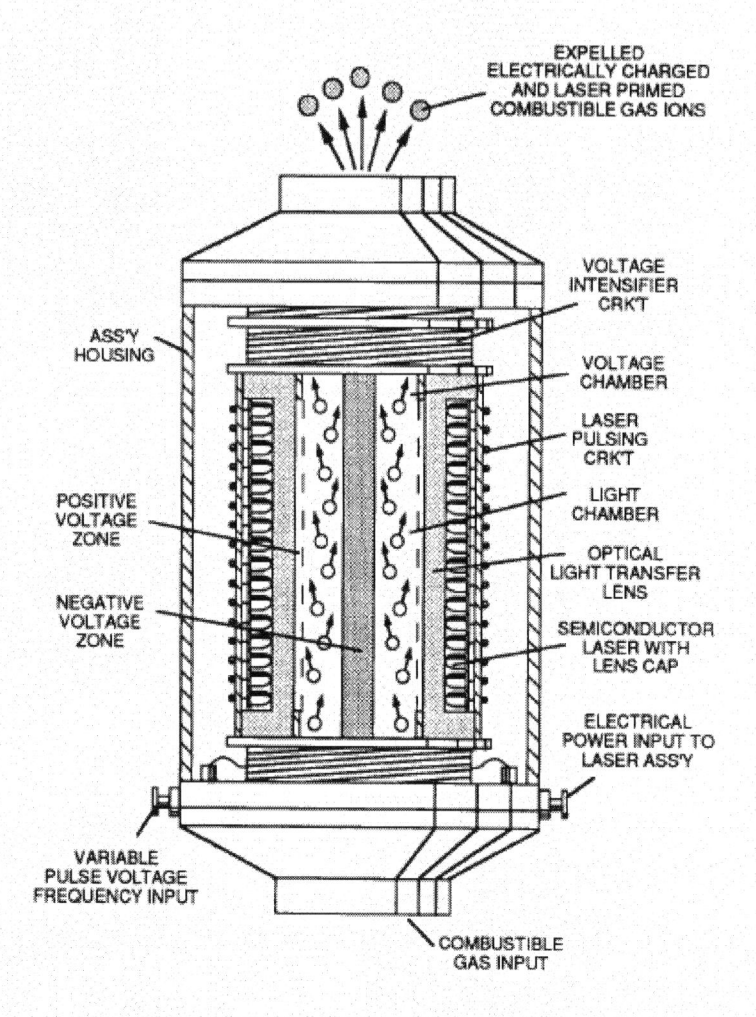

FIGURE 1-17: GAS RESONANT CAVITY

Stanley A. Meyer 1 - 21

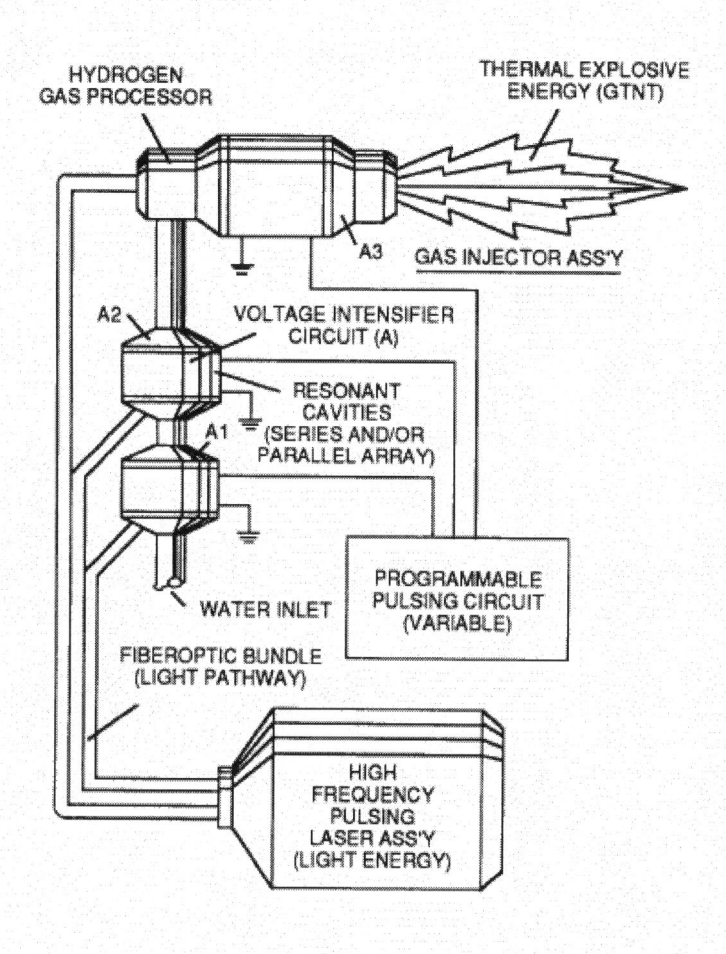

FIGURE 1-18: GAS INJECTOR FUEL CELL

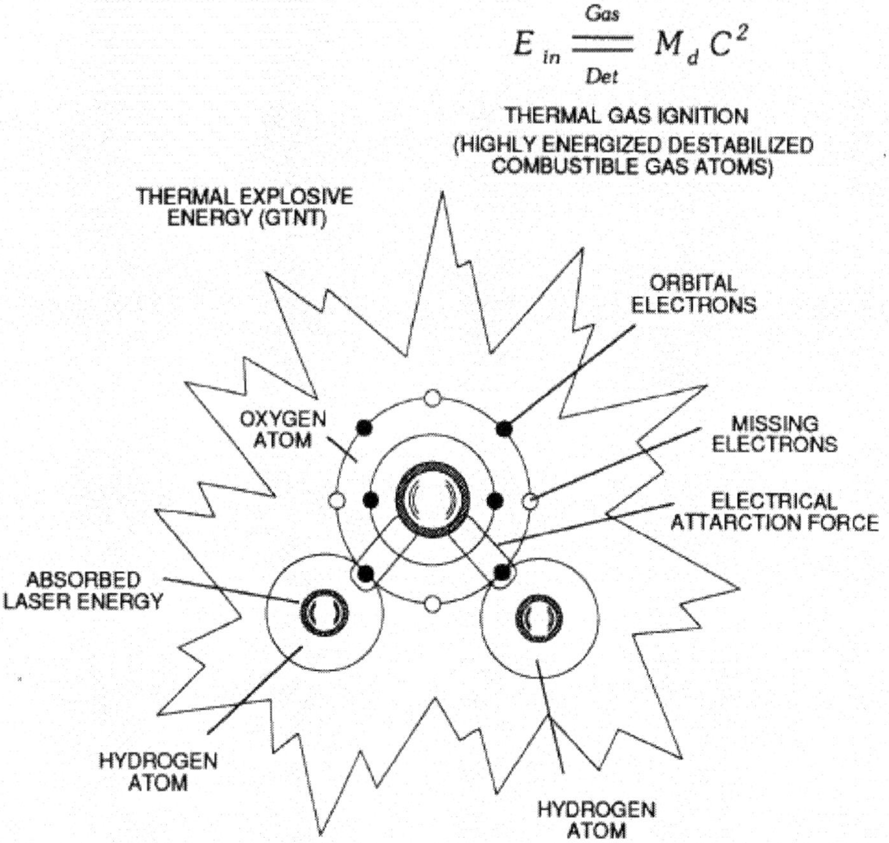

RE: WFC Hydrogen Fracturing Process   Memo WFC 420

$$E_{in} \underset{Det}{\overset{Gas}{=\!=\!=}} M_d C^2$$

THERMAL GAS IGNITION
(HIGHLY ENERGIZED DESTABILIZED
COMBUSTIBLE GAS ATOMS)

THERMAL EXPLOSIVE ENERGY (GTNT)

ORBITAL ELECTRONS

OXYGEN ATOM

MISSING ELECTRONS

ELECTRICAL ATTARCTION FORCE

ABSORBED LASER ENERGY

HYDROGEN ATOM

HYDROGEN ATOM

FIGURE 1-19: HYDROGEN FRACTURING PROCESS

Stanley A. Meyer

RE: WFC Hydrogen Fracturing Process — Memo WFC 420

$$E_{in} \underset{Det}{\overset{Gas}{=\!=\!=}} M_d C^2$$

THERMAL EXPLOSIVE ENERGY (gtnt)

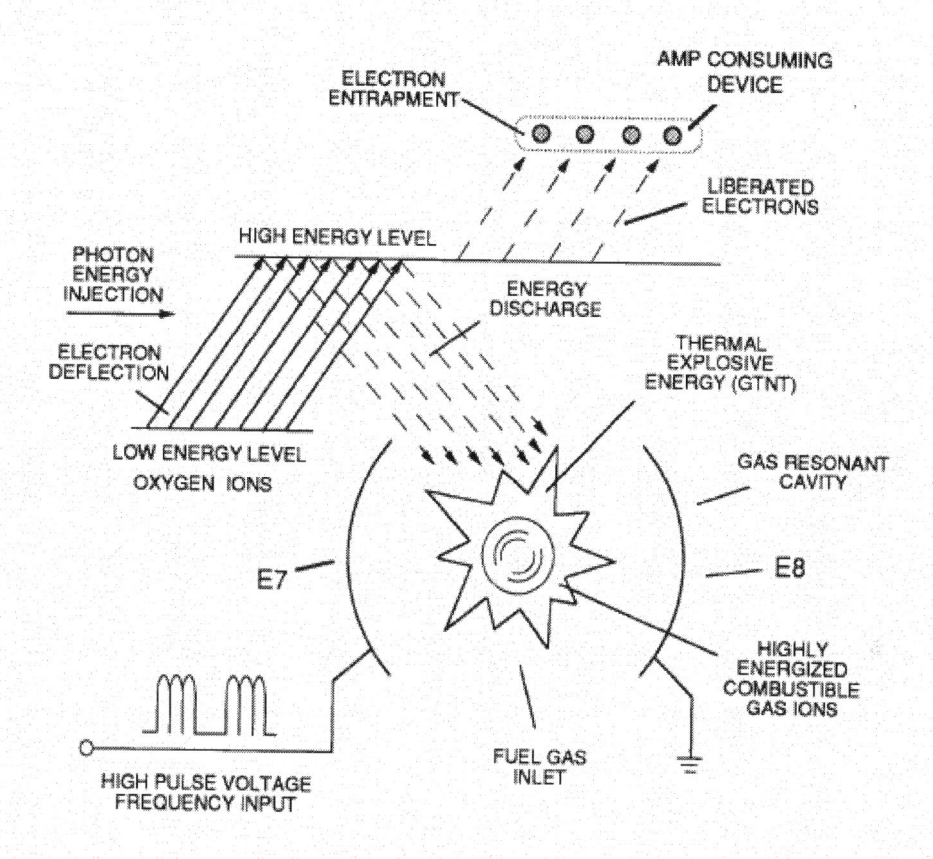

FIGURE 1-20: HYDROGEN GAS PROCESSOR

RE: WFC Hydrogen Fracturing Process　　　　　　　　　　　　　　　　Memo WFC 420

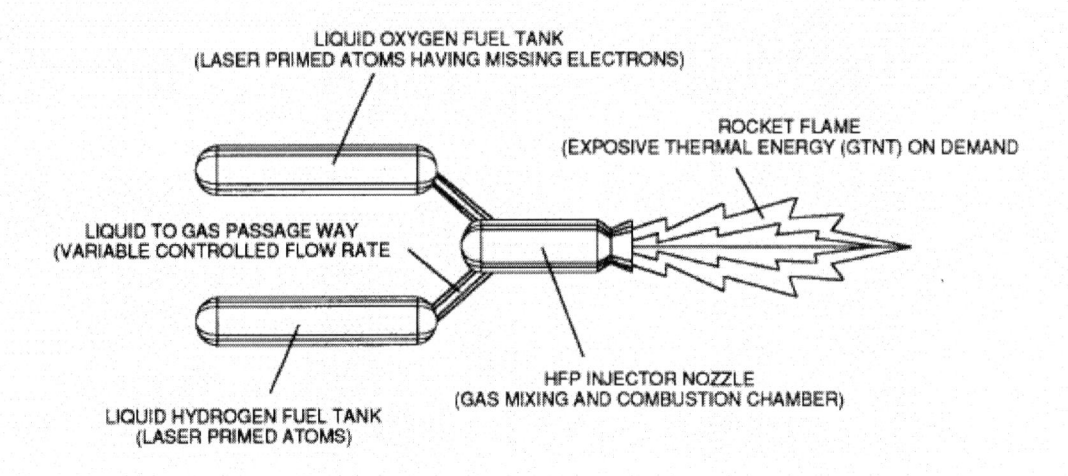

FIGURE 1-21: ATOMIC POWERED ROCKET ENGINE

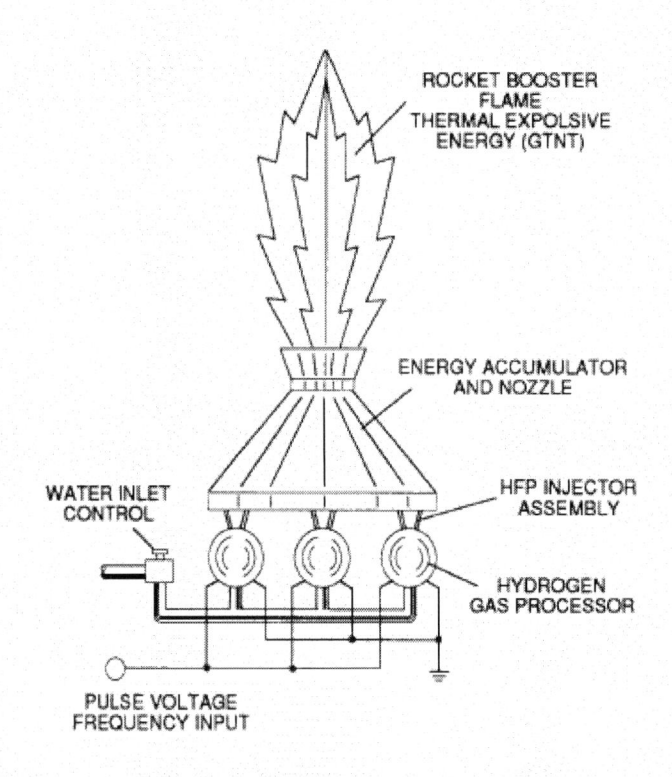

FIGURE 1-22: BOOSTER ROCKET ENGINE

Memo WFC 421

WATER FUEL CELL

Quenching Circuit Technology

Rendering Hydrogen Safer Than Natural Gas

The Quenching Circuit Technology is a combination and integration of several Gas-Processes that uses non-combustible gases to render hydrogen safer than Natural Gas.

The "Non-Burnable" gases are used to adjust hydrogen "Burn-Rate" to Fuel-Gas burning levels...recycled to stabilize Gas-Flame temperatures ...intermixed to sustain and maintain an hydrogen Gas-Flame... and used to prevent Spark-Ignition of supply gases.

The utilization and recycling of the non-combustible gases allows the Water Fuel Cell to become a Retrofit Energy System.

The Quenching Circuit Technology is systematically activated and performed in the following way:

Section 2

RE: Quenching Circuit Technology   Memo WFC 421

Quenching Circuit Technology

(Rendering Hydrogen safer than Natural Gas)

Operational Parameters

Spark-Ignition Tube

Spark-Ignition Tube (B) is a tubular test apparatus (1/8 diameter) that determines and measures the "Burn-Rate" of different types of Burnable Gases intermixed with Ambient Air, as illustrated in Figure (2-1).

Spark-Ignitor (A) causes and starts the Burnable Gas-Mixture (B) to undergo Gas-Ignition which, in turns, supports and allows Gas Combustion to take place...forming and sustaining a Gas-Flame. The expanding and moving Gas-Flame travels (away from spark-ignitor) the linear length of the gas filled tube (C) and is "detected" and "measured" (length between spark-ignitor and light-detector) in one second after gas-ignition. The Gas-Ignition Process, now, establishes the "Burn-Rate" of a Burnable Gas-Mixture in centimeters per second (cm/sec.), as illustrated in Figure (2-2).

Different types of "Burnable" Gas-Mixtures exposed to the Gas-Ignition Process were tested, measured, recorded and systematically arranged as to cm/sec. length, see vertical bar Graph (2-2) again. The Gas-Ignition Process was performed several times to establish the "average" Burn-Rate of the Fuel-Gases which, in turn, establishes the length of the vertical bars.

Gas Injection Process

Injecting and intermixing an Non-Combustible Gas (D) (non-burnable gas) with the "Burnable" Gas-Mixture (B) "changes" or "alters" the gas-mixture "Burn-Rate". Increasing the volume-amount of Non-Combustible Gas (D) diminishes and/or lowers the "Burn-Rate" of the Gas-Mixture (B/D) still further. Progressive and controlled intermixing of the non-combustible gases (B/D) allowed the "Burn-Rate" of Hydrogen to be "lowered" or "adjusted" to "match" or "co-equal" the "Burn-Rate" of other Fuel-Gases, see curve line in Figure (2-2).

In terms of operational performance, the Non-Burnable gas (D) does "Not" support the Gas Combustion Process since the Non-Burnable Gas (D) "restricts" or "retards" the speed at

RE: Quenching Circuit Technology   Memo WFC 421

---

which the Oxygen Atom unites with Hydrogen Atoms to cause Gas Combustion. The "Gas Retarding Process" is, of course, applicable to any type or combination of Burnable Gases or Burnable gas-mixture.

Gas Mixing Regulator

Inherently, the Water Fuel Cell allows the "Burn-Rate" of Hydrogen to be "Changed" or "adjusted" from 325 cm/sec. to 42 cm/sec. (Co-equalling Natural Gas Burning levels) since Non-Combustible Gases (such as Nitrogen, Argon, and other non-burnable gases) derived from Ambient Air dissolved in natural water performs the Gas Retarding Process...sustaining and maintaining an Open-Air Flame beyond 5,000-degrees F, as illustrated in Figure (2-3).

Natural water acts and performs as a "Gas-Mixing Regulator" when the Fuel-Cell is electrically energized by way of voltage stimulation (Electrical Polarization Process)...producing a uniform gas-mixture (B/D) regardless of the Gas Flow-Rate of the Fuel-Cell...producing a uniform gas-mixture (B/D) only when needed. In quiescent-state, the supply of gases (B/D) being released from the water bath is "terminated" and "stopped" when the Fuel-Cell becomes "de-energized". The unused water, of course, remains as a non-burnable liquid. The gases (B/D) above the water bath is "vented" for safety purposes.

Flame Temperature Adjustment

By capturing and recycling the expelled non-combustible gas (D) (derived from and supplied by the water bath) back into the sustained hydrogen gas-flame or Fuel-Cell causes the gas-flame temperature to be "changed" or "altered" by way of the Gas Retarding Process, as illustrated in Figure (2-4) as to Figure (2-3). The recycling gases (D) controlled by an Gas Flow-Regulator allows the gas flame-temperature to be "adjusted" or "calibrated" to any gas burning level (S), as so illustrated in Figure (2-2).

The "newly" formed and established gas flame-temperature remains constant regardless of the gas flow-rate of the Fuel-Cell. Continual feedback of non-combustible gases (D) is, hereinafter, called "The Gas Combustion Stabilization Process".

Automatically, the Gas Combustion Stabilization Process changes the "Burn-Rate" of the Fuel Cell gases (B/D) when obtaining the desired gas-flame temperature.

Stanley A. Meyer

RE: Quenching Circuit Technology   Memo WFC 421

Quenching Circuit

Spark-Ignition of the Fuel-Cell gases (B/D) is prevented when the "Gas Retarding Process" is used in conjunction with a "Quenching Circuit", as illustrated in Figure (2-3), (2-4), (2-5) and 2-6).

The non-combustible gases (D) separates and prevents the hydrogen atoms to unite with oxygen atoms to "bring-on" or "initiate" Gas-Ignition. The narrow passaway (at least 1/8 inch long and having a .015 diameter) prevents the moving gas atoms from "Re-Grouping". The alignment of the Fuel-Cell gases (B/D) inside the tubular-passaway is, hereinafter, called "The Quenching Circuit". The Quenching Circuit "Anti-Spark technique" is "independent" of both Gas-Velocity and Gas-Pressure.

Quenching Nozzle

Additional Quenching Circuits arranged in a Disc-shape configuration forms a "Quenching Nozzle" when attached to an "Quenching Tube", as illustrated in Figure (2-4) as to Figure (2-6). The Multi Gas-Port Disc compensates for increased Gas-Velocity while "preventing" spark-ignition of the Fuel-Cell gases. The overlapping Flame-Pattern re-ignites the expelling hydrogen gas-mixture (B/D) should Flame-Out occur. Ceramic material is used to form the "Quenching Disc" to "prevent" hole-size enlargement due to gas-oxidation.

The non-combustible gases (D) keeps the Ceramic Material "cool-to-the-touch" by projecting the Gas-Flame beyond and away from the disc-surface...the Quenching Disc remains "cool" even if the Gas-Flame Temperature exceeds the melting-point of the disc-material.

Quenching Tube

The Quenching Disc is extended into a Flexible Tube to transport the Fuel-Cell gases safely over long distances, as illustrated in Figure (2-7). The Spark-Arresting Gas-Line is, hereinafter, called "The Quenching Tube."

Catalytic Block Assembly

An Inverted hemispherical cavity placed on top of and in space relationship to the "Quenching Disc" insures total gas-combustion by recycling any "escaped" or "unused" burnable gases back into the gas-flame for Gas-Ignition...preventing Gas-Oxide formation, as illustrated in Figure (2-8) as to Figure (2-4).

Internal Combustion Engine

The Gas Combustion Stabilization Process (recycling non-combustible gases) is also applicable to operating an Internal Combustion Engine without changing Engine-Parts since the Gas Retarding Process allows the hydrogen "Burn-Rate to "equal" the "Burn-Rate" of Gasoline or Diesel-Fuel, as illustrated in Figure (2-2). The engine provides its own non-combustible gases derived from Ambient Air undergoing the gas-combustion process. Engine temperature remains the same since The Gas Stabilization Process is used.

Gas Grid System

Ambient Air is the prime source of Non-Combustible Gases when the Air-Gases are exposed to and passes through an Open-Air Flame, as illustrated in Figure (2-10). The Gas Combustion Process of the Gas-Flame eliminates oxygen and burnable gas atoms from the expelling gases...producing an endless supply of non-combustible gases.
Mixing the "processed" Air-Gases with an Hydrogen Supply Source sets up The Gas Retarding Process...allowing the Hydrogen Gas-Mixture to be transported safely through existing Gas-Grid System.

Operational Parameters

The utilization and recycling of non-combustible gases, now, renders hydrogen gas as safe as Natural Gas or any other Fuel-Gas...allowing the Water Fuel Cell to become a Retrofit Energy System.

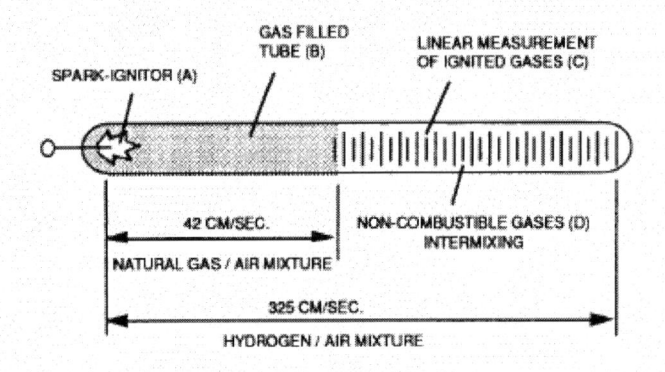

RE: WFC Quenching Circuit Technology   Memo WFC 421

FIGURE 2-1: SPARK IGNITION TUBE

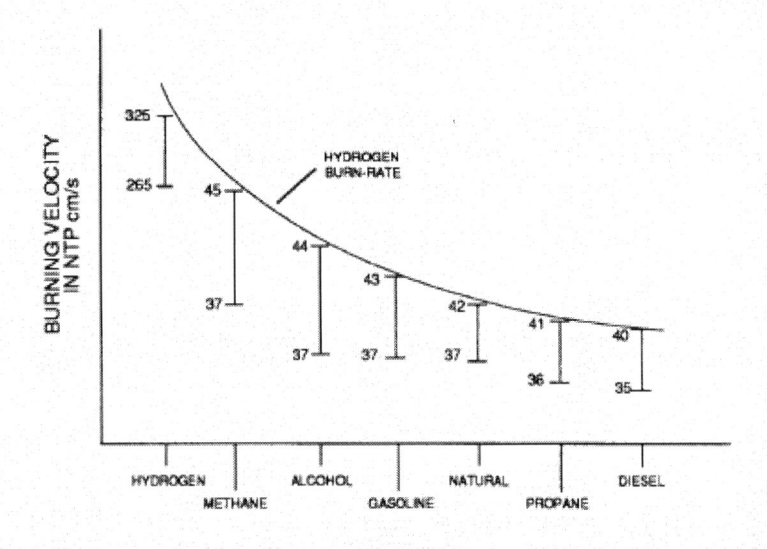

FIGURE 2-2: HYDROGEN BURN RATE

RE: WFC Quenching Circuit Technology

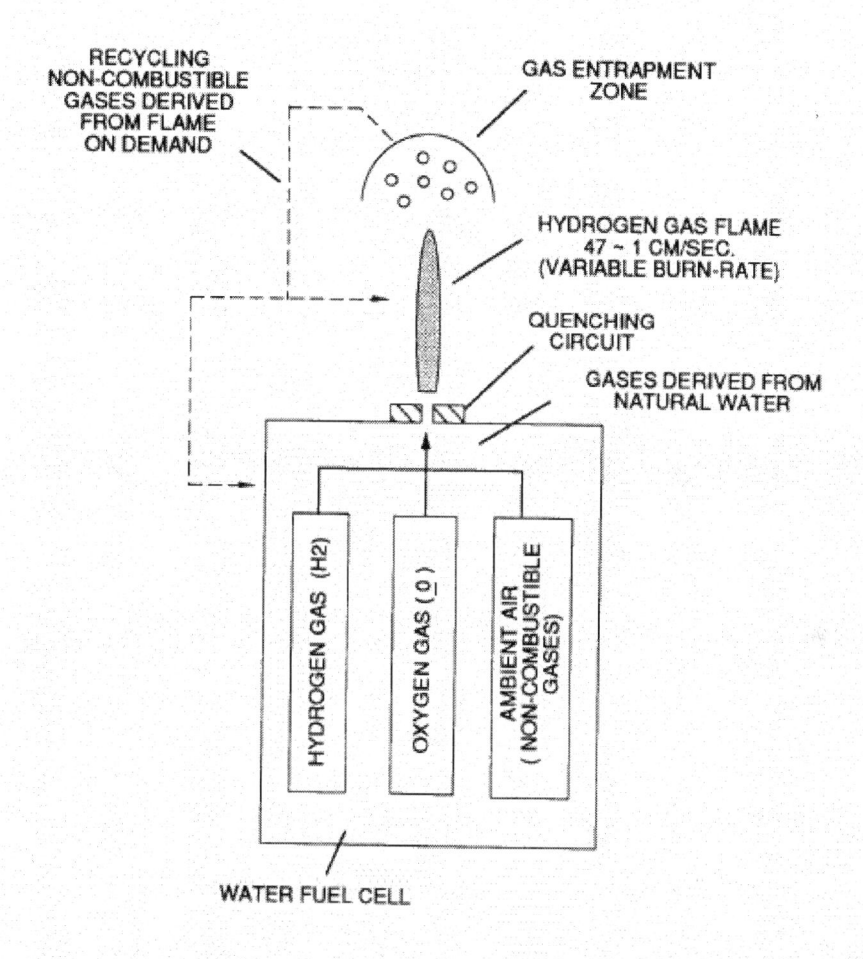

Memo WFC 421

FIGURE 2-3: GAS MIXING REGULATOR

Stanley A. Meyer

Page 2-6

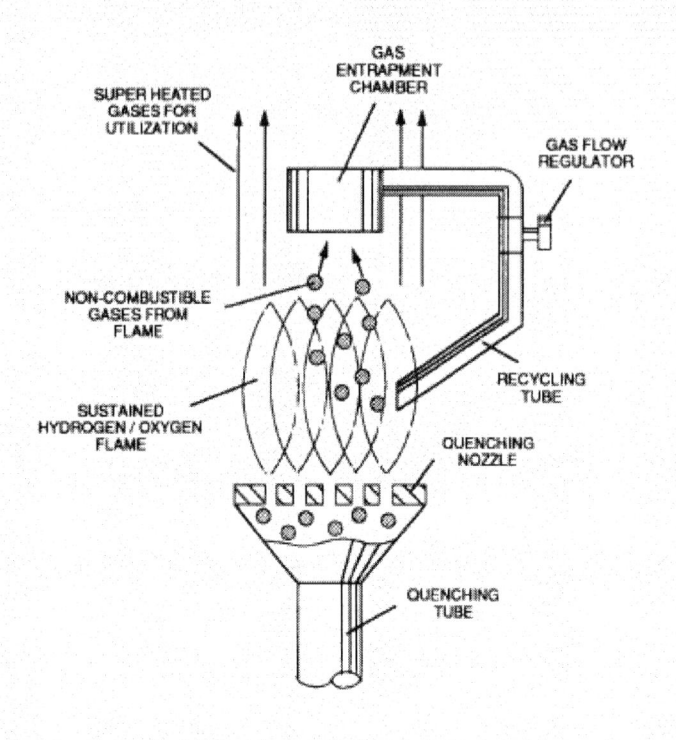

RE:WFC Quenching Circuit Technology   Memo WFC 421

FIGURE 2-4: ADJUSTING FLAME TEMPERATURE

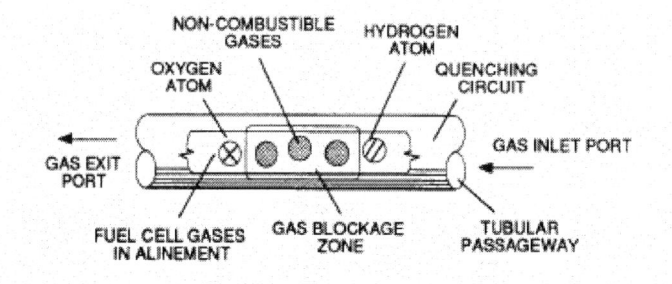

NOTE:
1) OXYGEN ATOM MUST UNITE WITH HYDROGEN ATOMS TO CAUSE GAS IGNITION.
2) TUBULAR PASSAGEWAY PREVENTS MOVING GAS ATOMS FROM REGROUPING.

FIGURE 2-5: PREVENTING GAS IGNITION

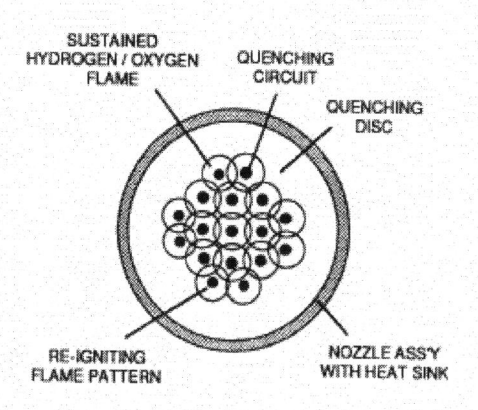

FIGURE 2-6: QUENCHING NOZZLE

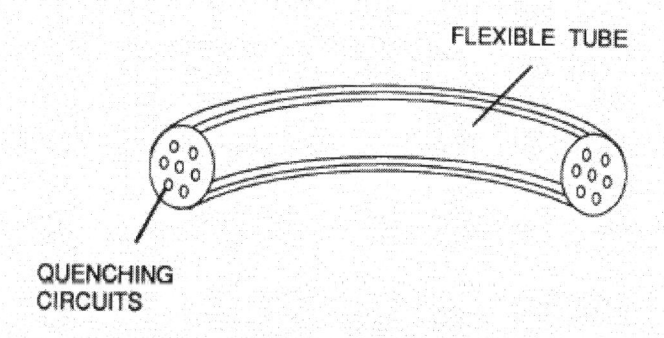

FIGURE 2-7: QUENCHING TUBE ALLOWS HYDROGEN TO BE DISTRIBUTED WITHOUT SPARK-IGNITION

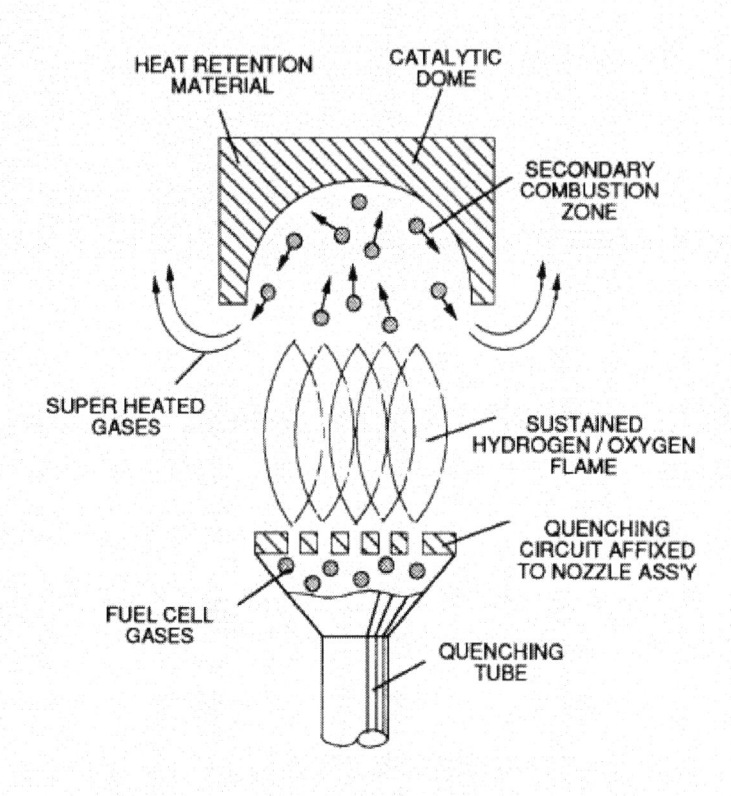

RE: WFC Quenching Circuit Technology   Memo WFC 421

FIGURE 2-8: CATALYTIC BLOCK ASS'Y

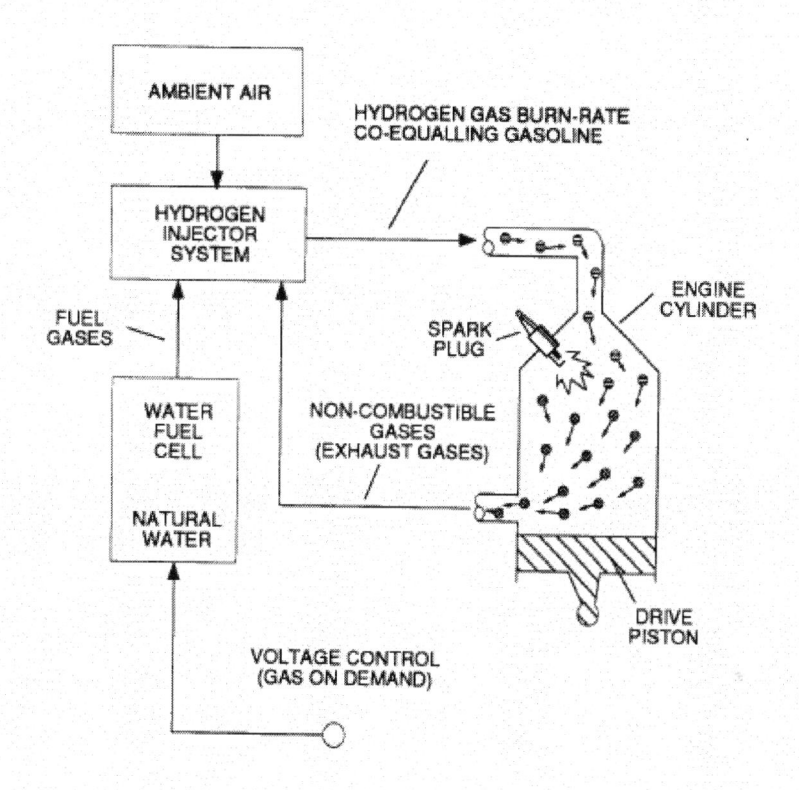

RE: WFC Quenching Circuit Technology   Memo WFC 421

FIGURE 2-9: WATER FUEL CELL RETROFITTED TO
A INTERNAL COMBUSTION ENGINE

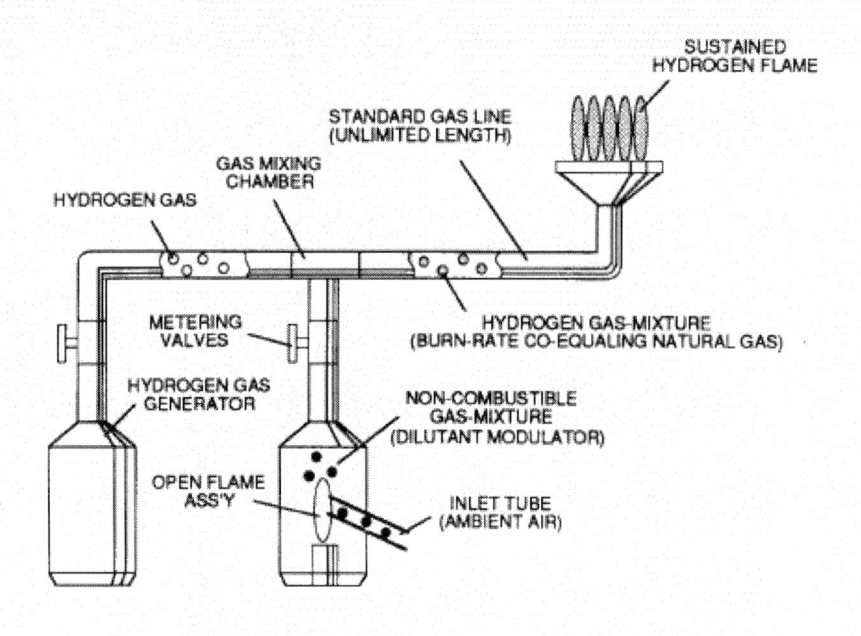

FIGURE 2-10: UTILIZING STANDARD GAS LINE TO TRANSPORT HYDROGEN GAS SAFER THAN NATURAL GAS

320

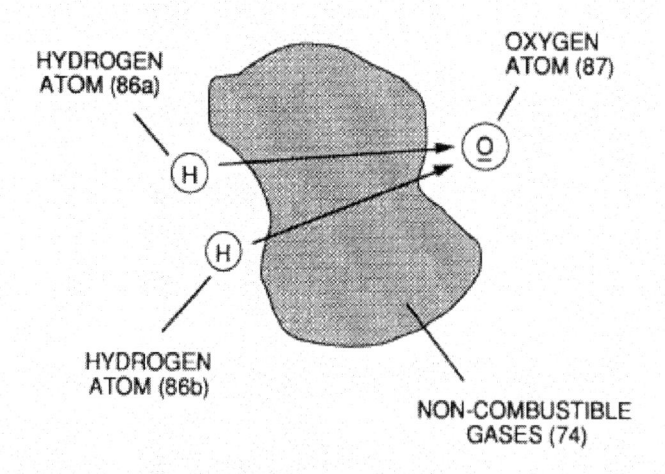

FIGURE 2-11: GAS MODULATOR

Memo WFC 422 DA

Water Fuel Cell

WFC Hydrogen Gas Management System

Water Fuel-Gas Injection System ®

The WFC Hydrogen Gas Management System encompasses many processing patents into a full system-engineering approach to help give a viable answer to the energy problem ... by using "Water" as a new Fuel - Source.

The WFC Hydrogen Gas Management (GMS) System not only economically produces "Hydrogen - Fuel" on demand from water but, also, renders Hydrogen - Fuel safer than natural gas ... allowing cars, trucks, or even jets to run on or be powered by water.

The WFC Hydrogen Gas Management (GMS) System is systematically activated and performed in the following ways as a retrofit energy - system ... defining " Mode of Operability" on how to use "Water" as Fuel.

Section 3

Memo WFC 422 DA

WFC HYDROGEN GAS MANAGEMENT SYSTEM
Water Fuel-Gas Injection System ®

Laser Accelerator Assembly (20)

Laser Accelerator Circuit (10) of Figure (4) which is a component part of Laser Accelerator Assembly (20) of Figure (3-10) uses a GaAs infrared emitting diode (1) of figure (3-9) to trigger a SDP8611 Optoschmitt light receiver (2) of Figure (3-9) from quiescent state (output logic high...B+) (13) to on-state (the minimum irradiance that will switch the output low) which switches or triggers the Optoschmitt (2) output to ground state (zero volts) (12). The peak wavelength (3) of Figure (3-9) being transmitted from the infrared emitting diode (led) (1) to the Optoschmitt receiver (2) is typically (935 nm) and allows the Optoschmitt (2) clock frequency (the speed by which the Optoschmitt changes logic state) to be (100 khz). Optical lens (4) of Figure (3-10) redirects and focuses the transmitted light source (3) of Figure (3-9) (traveling infrared light waves) to the Optoschmitt (2) by passing the light source through a series of concentric lenses (4a xxx 4n) of Figure (3-10) which become progressively smaller from the outer peripheral lens surface (4a) to the inner lens surface (4n). The spatially concentric lenses (4a xxx 4n) of Figure (3-10) causes the beam angle of the light source to trigger the Optoschmitt (2) beyond the minimum irradiance that is needed to switch the Optoschmitt from quiescent state (high logic state / B+) to on-state (output changing to zero volts).

The Derate linearly of light intensity is approximately 1.25mW/ degree C above 25 degree C at a spatial distance of .500 inches between the two infrared devices (1)(2) of Figure (3-9) as to Figure (3-10). Transmitted light source (3) is turn-on when a electrical power source of 5 volts is applied to the led (1) through dropping resister (5) by way of voltage regulator (6) connected to the car electrical system (7). Together, the matched infrared devices (1)(2) with optical lens (4) forms optical circuit (8) of Figure (3-9). Grouping additional optical circuits (8a xxx 8n) in a in-line or linear arrangement, now, forms Led Pickup Circuit (10) of Figure (3-9), as shown in Figure assembly (20) of Figure (3-10).

To perform a switch-logic function, light - gate (9) of Figure (3-9) as to Figure (3-10) is inserted between the matched infrared devices (1)(2) and moved in a linear displacement from one optical circuit (8x) to another optical circuit (8xx), as illustrated in Figure (3-9)(3-10) as to Figure (3-7). Once light-gate (9) blocks and prevents traveling light-beam (3) from reaching the matched Optoschmitt (8xx), the darken Optoschmitt (11) (non-energized) changes output state since the irradiance energy level (3) is reduced to, or below the release point...triggering opposite logic state (12). As light-gate (9) advances to the next optical circuit (8xxx) a new and separate low-state logic function (12) occurs while the previous optical circuit (8xx) reverts back to high-state logic (13). Advancing light-gate (9) still further performs the same opposite (alternate) logic-state switching in

a sequential manner until the advancing light-gate (9) reaches the last optical circuit (8n). Reversing the movement of light gate (9) performs the same high to low logic switch-function but in reverse sequential order. Reversing the direction of the light-gate (9) once again reinstates the original sequential switching order, as illustrated in Figure (3-7) and Figure (3-9).

Longevity and reliability of component life is typically 100,000 hours since led pickup circuit (10) of figure (3-9) utilizes no mechanical contacts to perform the sequential logic switch function. Light-gate (9) integrated with led pickup circuit (10) make up Laser Accelerator assembly (20), as shown in Figure (3-10). Light-gate (9) of Figure (3-10) is mechanically linked to the car acceleration pedal by way of cabling hookup (22).

Opposite placement of the matched infrared devices (1)(2) prevents bogus or false triggering of "low" logic state (12) during light-gate displacement (9a xxx 9n) of Figure (6)(7) and (8). If light emitting diodes (led) (1a xxx 1n) of figure (8) are electrically disconnected from D.C. power supply (6), then Led Pickup Circuit (10) outputs are switch to "low" logic state (12a xxx 12n) which disallows "low" logic state signal (12), resulting in a "shut-down" condition to Hydrogen Gas Control Circuit (200) of Figure (3-1). Disconnection of power supply (6) to Optoschmitt array (2a xxx 2n) of Figure (3-9) results in a similar "shut down" condition to control circuit (200), as further shown in Figure (3-1). This "shut-down" or "Switch-off" condition helps provide a fail-safe operable Fuel Cell (120) of Figure (3-20) by negating acceleration beyond driver's control.

Acceleration Control Circuit (30)

Moving light-gate (9) of figure (3-9) in direct relationship to the physical placement of optical circuits (8a xxx 8n), sets up a time variable (14a xxx 14n) of Figure (3-7) from optical circuits (8x) to another optical circuit (8xx) and/ or (8xxx) or to (8n) since the triggered low logic state (12) of Figure (3-7) and (3-8) moves in direct relationship to the displaced light-gate (9), as illustrated in Figure (3-12). Deflecting (moving) the light-gate (9) to position (8n) takes longer in response-time (14n) than deflecting the light-gate to position (8x) and/or (8xx) or (8xxxx). This variable response-time (14axx...12...xx14n) or signal output (15) of Figure (3-5) is, now, electrically transmitted to Acceleration Control Circuit (30) of Figure (3-5) since Laser Accelerator Assembly (20) of figure (3-10) converts mechanical displacement (9a xxx 9n) to electrical time-

RE: WFC Hydrogen Gas Management System   Memo WFC 422 DA

response (14a xxx 14n) of Figure (3-7) by linearly moving (forward and/or reverse direction) "low" logic state signal (12) in a array of "high" logic state output signals (13a xxx 13n), as further illustrated in Figure (3-8) and Figure (3-12). In some cases reverse signal-logic (12a xxx...13...xx12n) is applicable by using SDP 8601 Optoschmitt which switches logic state from Quiescent state ("low" to "high" logic state) when de-energized.

Since Led Pickup Circuit (10) of Figure (3-9) operates up to 100 Khz range or above, electrical sensitivity of Opto-circuit (8) provides a instantaneous response to Driver's acceleration, de-acceleration, or cruise control demands.

As signal output (15) of figure (4) (14a xxx ...12...xx14n) is being received by acceleration control circuit (30) of Figure (3-5) as to Figure (3-12), circuit (30) converts incoming time-response signal (14a xxx...12...xx14n) into a variable time-base unipolar pulse (16), as shown in Figure (3-8). Circuit (30) electronically and automatically scans output signal-array (14a xxx...12...xx14n) (15) until circuit (30) locates, momentarily registers, and translates response-time (14a xxx...12) into a variable unipolar pulse (17/18) of Figure (3-8). The sweeping action of the scanning circuit (30) always starts from position (9a) and moves point (8ax) to point (8axxx) of Figure (3-9) (3-12) until logic-point (12) is detected. Once logic signal (12) is detected, the sweeping action toggles and recycles back to start-position (9a). This toggling (flip back) action electronically determines variable time-response (14a xxx) regardless of wherever logic point (12) is being momentarily displaced within circuit array (13a xxx 13n).Toggling action at full-scale deflection (13a xxx 13n) occurs in the range of (10) Khz or above and thus, allows instant response to Driver's acceleration demands.

Toggling-time (scanning -time) is directly synchronized to light gate (9) displacement which, in turns, circuit (30) further sets up and establishes a given pulse shape (16) of Figure (3-8). Circuit (30) continues to increase pulse width (17axxxx) of Figure (3-8) as the monitored (detected by scanning) toggling-time (14a xxxx...12)) increases when logic-point (12) moves farther away from start-position (9a) to stop-position (9n), as further shown in Figure (3-13) as to Figure (3-12). Pulse width (17a xxx 17n) diminishes when logic-point (12) reverses direction to start-position (9a). Finally, circuit (30) reproduces the variable controlled pulse-shape (16) in a continuous repetitive manner (16a xxx 16n) of Figure (3-13) and electrically transmits the resultant pulse-train signal (19) to Analog Voltage Circuit (40), as shown in Figure (3-5).

In retrospect to engine performance (gas pedal attenuation) (21) of Figure (3-10), a wider pulse width (17a xxx) of Figure (3-13C) increases (accelerates) engine R.P.M.; whereas, smaller pulse-width (17ax) reduces (de-accelerates) engine R.P.M.. Cruising speed (3-13B) of Figure (3-13) is simply accomplished when pulse width remains constant.

Stanley A. Meyer

RE: WFC Hydrogen Gas Management System   Memo WFC 422 DA

Incoming clock pulse (21a xxx 21n) of Figure (3-16) originating from Pulse Frequency Generator (70) of Figure (3-5) sets up the scan-rate (toggling) by which signal input (15) of Figure (3-5) is electronically scanned by circuit (30). The resultant clock pulse (21) of Figure (3-16) as to Figure (3-5) is always adjusted to exceed driver's response time to allow for instant acceleration control.

Analog Voltage generator (40)

The generated digital signal (19) being electrically transmitted from accelerated control circuit (30) of Figure (3-5) is, now, electronically detected, translated, and converted into a analog voltage signal (22) which is continuously proportionate to input signal (19) by analog voltage Generator Circuit (40) of Figure (3-5). The newly formed analog signal (22) of Figure (3-14) is a voltage level signal that varies continuously in both time and amplitude to produce a voltage level which is directly proportional to the physical change in pulse train (16a xxx 16n) of Figure (3-13). As pulse width (17ax) of signal (19) changes so does analog voltage level output (23) of Figure (3-14). Widening pulse width to stop-position (17a xxxx 17n) of Figure (3-13) causes analog signal (22) to increase to higher voltages levels; whereas, analog voltage level (22) drops (become lower in value) in voltage level when pulse width decreases to start-position (17a). The resultant and varied voltage level (22a xx) varies smoothly over a continuous range of voltage valves (22a xxx 22n) rather than in discrete steps, as illustrated in linear graph (23) of Figure (3-14).

In terms of functionalability and purpose, analog circuit (40) of Figure (3-5) provides a variable (controlled) voltage output (23) in direct relationship to light gate (9) displacement which, in turns, sets up and controls Resonant Action (160) of Figure (3-23) that produces Fuel Gases on demand. Analog circuit (40) also calibrates both engine idling speed (22ax) and maximum engine R.P.M. (22a xxx 22n) by adjusting and maintaining a predetermined or given low (24) and high voltage levels respectively, as further illustrated in Figure (3-14). Voltage valves or levels (22a xxx 22n) simply controls the applied voltage potential across Resonant Cavity Assembly (120) of Figure (3-22) through voltage amplitude control circuit (50) of Figure (3-5) which is is electrically linked to primary coil (26) of Figure (3-21) of Voltage Intensifier Circuit (60) of Figure (3-5).

Voltage Amplitude Control Circuit (50)

Voltage amplitude control circuit (50) of Figure (3-5) performs several functions simultaneously: First, regulates car battery electrical voltage potential (32) of Figure (3-15) being applied to primary coil (26) of Figure (3-21); and secondly, regulates gas pressure of Fuel Cell (120) of Figure (3-22), as graphically depicted in Figure (3-15). Each regulatory stage (27) and (28) works separately and independent of each other but are electronically linked or coupled together to produce a common analog signal (32) having a predetermined voltage level (32a xxx), as further shown in Figure (3-15).

Regulator stage (27) of circuit (50) converts battery voltage potential (29) of Figure (3-6) via electrical terminal (31) of Figure (3-5) as to Figure (3-6) into a analog voltage signal (32) of Figure (3-15) which corresponds to but is electrically isolated (crossover voltage from two separate power supplies) from incoming gas volume signal (23) of Figure (3-14), as shown in Figure (3-5). Variable voltage range (32a xxx 32n) from one (1) up to twelve (12) volts (regulating battery voltage) is applied across primary coil (26) of Voltage Intensifier Circuit (60) of Figure (3-21). Second regulator stage (28) simply acts and function as a gas regulator (33) by preventing Fuel Gas production beyond a predetermined gas pressure level (34) of Figure (3-15) during Fuel Cell operations and, as such, maintains constant gas pressure to Fuel Injectors (36) of Figure (3-1) regardless of engine performance (R.P.M. response). If for example, Fuel Gas production is greater than demand, then, analog signal (32) is reduced to proper voltage level (35) (voltage level directly determines gas pressure via Resonant Action) required to maintain gas pressure (34), Conversely, analog signal (32) is always allowed to exceed voltage level (35) during injection (36) of Figure (3-1) until gas-point (34) is reached. In cases where linear voltage (32) drops (descending value) below gas-point (35) then gas regulator stage (28) increases voltage amplitude (32a xxx 32n) (analog voltage) to voltage point (35). If gas pressure (34a xx) should exceed gas point (35) during injector off-time, gas pressure release valve (75) of Figure (3-24) (gas venting 37 of Figure 3-15) expels Fuel gases (88) until gas point (34) is either reach or a delay timing circuit activates Safety Control Circuit (14) of Figure (3-6) which, in turns, switches off or disconnects applied electrical power (28) to Fuel Cell electrical system (400) of Figure (3-6).

Gas logic circuit (310) of Figure (3-5) supplies logic function to Voltage amplitude control circuit (50) to maintain proper gas pressure to gas injector (36) of Figure (3-1) by electronically monitoring achieved gas pressure via pressure sensor (73) of Figure (3-24).

In terms of operability, Laser Accelerator Assembly (20) of Figure (3-5) is, now, attenuating battery voltage potential (32a xxx 32n) which is electrically connected to voltage Intensifier Circuit (60) of Figure (3-5).

RE: WFC Hydrogen Gas Management System   Memo WFC 422 DA

Variable Pulse Frequency Generator (70)

Circuit (70) of Figure (3-5) is a multi pulse-frequency generator which produces several clock pulses (simultaneously) having different pulse-frequency but maintaining a 50% duty cycle pulse (39) configuration, as illustrated in Figure (3-16). Pulse on-time (37) and pulse off-time (38) are equally displaced to form duty pulse (39) which is duplicated in succession to produce pulse train (41) of Figure (3-16). Increasing the number of duty pulses (39a xxx 39n) up to pulse frequency range of 10Khz or above now forms clock signal (21) of Figure (3-5) which, in turns, performs the scanning function of Acceleration Control Circuit (30) of Figure (3-5). Circuit (70) also produces another independent and separate clock signal (41a xxx 41n) which is electrically transmitted to and become incoming clock signal (42) for Gated Pulse Frequency Generator Circuit (80) of Figure (3-5). In both cases, pulse frequency range of each clock signal (21) and (42) can be altered or change (controlled independent of each other) to obtain peak performance of Fuel Cell System (100) of Figure (3-5).

Gated Pulse Frequency generator (80)

Gated Pulse Circuit (80) of Figure (3-5) switches "off" and "on" sections of incoming clock signal (42) to form gated pulse (45) which is, in turn, duplicated in succession to produce gated pulse train (46a xxx 46n) of Figure (3-17). Together pulse train (44a xxx 44n) and pulse off-time (43) forms gated pulse duty cycle (45). Pulse train (44a xxx 44n) is exactly the same as pulse train (41a xxx 41n) and its established pulse frequency (number of pulse cycles per unit of time) changes uniformly when pulse generator (70) of Figure (3-5) is calibrated and adjusted for system operations.

Newly formed gated duty pulse (45) is proportional to the physical change in pulse train (44a xxx 44n) when circuit (80) is adjusted for calibration purposes. Pulse train (44a xxx 44n) becomes widen while pulse off-time width (43) becomes smaller, simultaneously. Conversely, opposite pulse shaping occurs when circuit (80) of Figure (3-5) is calibrated in reverse order.

Cell Driver Circuit (90)

In either case, the resultant or varied pulse train (47a xxx 47n) (calibration of 44a xxx 44n) becomes incoming gated pulse signal (48) of figure (3-5) to cell driver circuit (90) of Figure (3-5) which performs a switching function by switching "off" and "on" electric ground being applied to

RE: WFC Hydrogen Gas Management System   Memo WFC 422 DA

opposite side (48) of primary coil (26) of Figure (3-19). The resultant pulse wave form (49a xxx 49n) of Figure (3-18) superimposed onto primary coil (26) is exact duplicate of proportional pulse train (47a xxx 47n). However, each pulse train (47) (49) are electrically isolated from each other. Only voltage cross-over from regulated power supply (150) of Figure (3-6) to battery supply (28) occurs, as illustrated in Figure (3-6).

Voltage Intensifier Circuit (60)

By integrating and joining together variable voltage amplitude control signal (32a xxx 32n) of Figure (3-15) with variable controlled switch-gate (49a xxx 49n) of Figure (3-18) across primary coil (26) of Figure (3-22), variable amplitude pulse-train (51a xxx 51n) of Figure (3-19) is electromagnetically coupled (transformer action) to secondary coil (52) of Figure (3-22) by way of pulsing core (53) of Figure (3-23) as to Figure (3-22).

Analog voltage signal (32a xxx 32n) of Figure (3-15) allows pulse train (51a xxx 51n) voltage amplitude (Vo xxx Vn) of Figure (3-19) to vary from one up to twelve volts (battery supply 28 of Figure 3-6) by attenuating Laser Accelerator circuit (10) of Figure (3-5) via Hydrogen Gas Control Circuit (100). Variable pulse frequency generator (70) of Figure (3-5) varies and adjusts pulse frequency (63) (50% duty cycle pulse) while gated pulse frequency generator (80) of Figure (3-5) varies and adjusts pulse width (54a xxx 54n). These controlled and variable pulse features are, now, translated to Resonant Charging pulse train (65a xxx 65n) of Figure (3-21) via Unipolar pulse train (64a xxx 64n) of Figure (3-20) during Resonant Action (160) of Figure (3-26) when signal coupling is applied across Resonant Cavity (170) of Figure (3-24) via positive voltage zone (66).

Negative electrical voltage potential (61) of pulse wave (65a xxx 65n) of Figure (3-21) is simultaneously applied to negative voltage zone (67) via Resonant Charging Choke (62) of Figure (3-22) which is electrically linked to opposite end of Primary Coil (26). The resultant signal coupling (65a xx 65n) of Figure (3-21) is accomplished since primary coil (26), pulsing core (53), secondary coil (52), switching diode (55), resonant charging choke (56), resonant cavity assembly (170), natural water (68), and variable resonant charging choke (62) forms Voltage Intensifier Circuit (60) of Figure (3-22), as illustrated in Figure (3-22) as to Figure (3-23). Negative electrical ground (61) of Voltage Intensifier circuit (60) of Figure (3-22) is electrically isolated from primary electrical ground (48) of Figure (3-22).

Pulsing transformer (26/52) of Figure (3-22) steps up voltage amplitude or voltage potential (Vo xxx Vn) of Figure (3-19) during pulsing operations. Primary coil (26) is electrically

Stanley A. Meyer

isolated (no electrical connection between primary 26 and secondary coil 52) to form Voltage Intensifier Circuit (60) of Figure (3-22). Voltage amplitude or voltage potential (Vo xxx Vn) is increased when secondary coil (52) is wrapped with more turns of wire. Isolated electrical ground (61) prevents electron flow from input circuit ground (48).

Switching diode (55) of Figure (3-22) not only acts as a blocking diode by preventing electrical "shorting" to secondary coil (52) during pulse off-time (69) of Figure (3-20) since diode (55) "only" conducts electrical energy in the direction of schematic arrow; but, also, and at the same time functions as a electronic switch which opens electrical circuit (60) during pulse off-time...allowing magnetic fields of both inductor coils (56/57) to collapse...forming pulse train (64a xxx 64n).

Resonant charging choke (56) in series with Excitor-Array (160) of Figure (25) forms an inductor-capacitor circuit (180) of Figure (3-28) since Excitor-Array (66/67) acts and performs as an capacitor (dielectric liquid between opposite electrical plates) during pulsing operations. The dielectric properties (insulator to the flow of amps) of natural water (68) of Figure (3-28) as to Figure (3-26) (dielectric constant of water being 78.54 @ 20C in 1 atm pressure) between electrical plates (66/67) forms capacitor (57), as illustrated in (170) of Figure (3-25). Water now becomes part of Voltage Intensifier circuit in the form of "resistance" between electrical ground (67) and pulse-frequency positive potential (66)...helping to prevent electron flow within pulsing circuit (60) of Figure (3-22).

Inductor (56) and capacitor (57) properties of LC circuit (180) is therefore "tuned" to resonant at a given frequency. Resonant frequency (63) of Figure (3-19) can be raised or lowered by changing the inductance (56) and/or capacitance (57) valves. The established resonant frequency is, of course, independent of voltage amplitude, as illustrated in Figure (3-21) as to Figure (3-18). The value of inductor (56), value of capacitor (57), and the pulse-frequency (63) of voltage (Vo xxx Vn) being applied across the LC circuit determined the impedance of LC circuit (Figure 3-28).

The impedance of inductor (56) and capacitor (57) in series, Z series is given by

(Eq 1)

$$Z\text{ series} = (X_c - X_l)$$

RE: WFC Hydrogen Gas Management System   Memo WFC 422 DA

where

(Eq 2) $$X_c = \frac{1}{2\pi fc}$$

(Eq 3) $$X_l = 2\pi fl$$

Resonant frequency (63) of LC circuit in series is given by

(Eq 4) $$F = \frac{1}{2\pi \sqrt{LC}}$$

Ohm's law of LC circuit (180) of Figure (3-28) in series is given by

(Eq 5) $$V_t = IZ$$

The voltage across inductor (56) or capacitor (57) is greater than applied voltage (49) of Figure (3-18). At frequency close to resonance, the voltage across the individual components is higher than applied voltage (49), and, at resonant frequency, the voltage (Vt) of Figure (3-28) across both inductor and the capacitor are theoretically infinite. However, physical constraints of components and circuit interaction prevents the voltage from reaching infinity.

The voltage (Vl) across inductor (56) is given by equation

(Eq 6) $$V_l = \frac{V_t X_l}{(X_l - X_c)}$$

Voltage (Vc) of Figure (3-28) across the capacitor is given by

(Eq 7) $$V_c = \frac{V_t X_c}{(X_l - X_c)}$$

During resonant interaction, the incoming unipolar pulse train (64a xxx 64n) of Figure (3-20) as to Figure (3-21) produces a step charging voltage effect across excitor-array (66/67) (57) as

RE: WFC Hydrogen Gas Management System   Memo WFC 422 DA

so illustrated in Figure (3-21). Voltage intensity increases from zero "ground-state" to an high positive voltage potential in an progressive function. Once voltage-pulse (64) is terminated or switch-off, voltage potential returns to "ground-state" (61) or near ground-state (diode 55 maintains voltage charged across capacitor 57) to start the voltage deflection process over again as pulse train (64a xxx 64n) continues to be duplicated.

Voltage intensity or level across excitor array (57) can exceed 20,000 volts due to circuit (60) interaction and is directly related to pulse train (64a xxx 64n) variable amplitude input.

Inductor (56) is made of or composed of resistive wire to further restricts D.C. current flow beyond inductance reaction (Xl), and, is given by

(Eq 8)
$$Z = \sqrt{R_1^2 + (X_l - X_c)^2}$$

Variable inductor-coil (62) of Figure (3-22), similar to inductor (56) connected to opposite polarity voltage zone (67) further inhibits electron movement or deflection within voltage intensifier circuit (60). Movable wiper arm (73) of Figure (3-22) fine "tunes" "resonant action" during pulsing operations. Inductor (62) in relationship to inductor (56) electrically balances the opposite electrical potential across voltage zone (66/67).

Since pickup coil (52) is also composed of or made of resistive wire-coil, then, total circuit resistance is given by

(Eq 9)
$$Z = R_1 + Z_2 + Z_3 + R_E$$

where, R_E is the dielectric constant of natural water.

Ohm's law as to applied electrical power, which is

(Eq 10)
$$E = IR$$

Stanley A. Meyer

RE: WFC Hydrogen Gas Management System Memo WFC 422 DA

where,

(Eq 11)
$$P = EI$$

Whereby,

electrical power (P) is an linear relationship between two variables, voltage (E) and amps (I).

Amp restriction beyond "resonant action" occurs when unipolar magnetic field coupling (71) of Figure (3-23) is allowed to simultaneously drop (pulsating magnetic field) across both resonant charging chokes (56/62) during pulsing operations since electron mass is a electromagnetic entity which is subject to inductor fields (56/62) produced by pulsating magnetic field (71a xxx 71n) of Figure (3-23). Amp leakage (electron coupling to water) to water bath (68) of Figure (3-24) is further prevented by encapsulating resonant cavity (57) in delrin material (72) of Figure (3-25) which is an electrical insulator to high voltage. Delrin material (72) insulator value remains intact since insulation material (72) is resilient to water absorption.

Inherently, then, pulsing core (53) of Figure (3-23) aids amp restriction while voltage intensifier circuit (190) is being "tuned" (adjusting pulse train 49a xxx 49n pulse-frequency 63 via pulse frequency generator 70 of figure 3-5) to match the resonant frequency properties of water bath (68) of Figure (3-22), as illustrated in Fuel Cell (120) of Figure (3-24). The resultant interfacing voltage circuit (190), now, exposes water molecule (210) of Figure (3-27) to a pulsating high intensity voltage field (65a xxx 65n) of opposite polarity (66/67) while restricting amp flow within circuit (60) of Figure (3-22).

Voltage Dynamics

Voltage is "electrical pressure" or "electrical force" within electrical circuit (60) and is known as voltage potential (65a xxx 65n) of Figure (3-21). The higher the voltage potential (Vo xxx Vn), the greater "electrical attraction force" (qq') or " electrical repelling force" (ww') of

RE: WFC Hydrogen Gas Management System   Memo WFC 422 DA

Figure (3-29) is applied to electrical circuit (60) of Figure (3-22). Voltage potential (65) is an "unaltered" or "unchanged" energy-state when "electron movement" or "electron deflection" is prevented or restricted within electrical circuit (190) of Figure (3-23).

Unlike voltage charges within electrical circuit (60) steps up "electrical attraction force" (qq'); whereas, like electrical charges within the same electrical circuit (60) encourages an "repelling action" (ww'), as illustrated in Figure (3-29). In both cases, electrical charge deflection or movement is directly related to applied voltage (65). These electrical "forces" are known as "voltage fields" and can exhibit either a positive (66) or negative (67) electrical charge.

Likewise, Ions or charged particles (atoms having missing or sharing electrons between unlike atoms) within electrical circuit (60) having unlike electrical charges are attracted to each other. Ions or particles mass having the same or like electrical charges will move away from one another, as illustrated in (220) of Figure (3-29).

Furthermore, electrical charged ions or particles can move toward stationary voltage fields or voltage zones (66/67) of opposite polarity, and, is given by Newton's second law

$$A = \frac{F}{M}$$ (Eq 12)

Where, the acceleration (A) of a particle mass (M) acted on by a net force (F).
Whereby, net force (F) is the "electrical attraction force" (qq') between opposite electrically charged entities (210) of Figure (3-27), and, is given by Coulomb's law

$$F = \frac{qq'}{R^2}$$ (Eq 13)

Whereas, difference of potential between two charges is measured by the work necessary to bring

RE: WFC Hydrogen Gas Management System　　　　　　　　　　Memo WFC 422 DA

the charges together, and, is given by

(Eq 14)
$$V = \frac{q}{e^R}$$

The potential at a point to a charge (q) at a distance (R) in a medium whose dielectric constant is (e).

Electrically Charged Water Molecule

Atomic structure of an atom (76) and (77) of Figure (3-27) exhibits two types of electrical charged mass entities, orbital electrons (79) having negative electrical charges (-) and a nucleus (84) (at least one proton) having a positive electrical charged (+). The positive electrical charge of the nucleus equals the sum total of all negative electrical charged electrons when the atom is in "stable-state." In stable state or normal-state, the number of electrons equals the number of protons to give the atom "no" net electrical charge.

Whenever one or more electrons are "dislodged" from the atom, the atom takes-on a net positive electrical charge and is called a positive ion. If a electron combines with a stable or normal atom, the atom has a net negative charge and is called a negative ion.

Voltage potential (65) within electrical circuit (60) can cause one or more electrons (79) to be dislodged from the water molecule atom (85) of Figure (3-26) due to opposite electrical polarity attraction (qq') of Figure (3-29) between unlike charged entities, as shown in (160) of Figure (3-26) as to Newton's and Coulomb's laws of electrical-force. These same laws of electrical-force (qq') is used to combine or join atoms together by way of covalent bonding (opposite electrical forces) to form a molecule of water (85), as illustrated in (210) of Figure (3-27).

The liquid molecule of water (210) of Figure (3-27) is formed when the two hydrogen atoms (77a/b) takes-on a net "positive electrical charge" (78), which is, equal to the net "negative electrical charge" (81) of the oxygen atom (76). The resultant electrical force (qq') between the opposite electrical charged hydrogen (77) and oxygen (76) atoms keeps water molecule (210) intact

RE: WFC Hydrogen Gas Management System   Memo WFC 422 DA

when the hydrogen atom (77) shares its electron (84) with oxygen atom (76). The electrical strength of attraction force (qq') between the water molecule atoms is determined by the electrical-size of the hydrogen atoms and the displacement of its negative charged electrons (84) during covalent sharing. Oxygen atom becomes negative electrical charged (81) since oxygen atom (76), now, has a total of ten negative charged electrons (79a xxx 79n) in its "K" plus "L" orbits while maintaining it's original eight positive charged protons which makes up oxygen nucleus (83). Since the hydrogen proton (84) (hydrogen nucleus) remain (after covalent link up), then the hydrogen atom takes-on a positive charged (78) co-equalling the positive charge of the hydrogen nucleus proton (84). Together, the total net charge of water molecule (85) is zero despite the fact that each water molecule atom retains its electrical charge. In other words, water molecule (85) is a electrically bipolar molecule having a stable configuration of charged atoms bound together by electrostatic force (qq'). Electromagnetic bonding forces between unlike atoms (76/77) are negligible or non-existence, since oxygen atom (76) electrons are paired together, while rotating in opposite direction which, in turn, causes oxygen atom (76) to be electromagnetically neutral to hydrogen atom (77). Electron theory of magnetism requires orbital electrons to spin in the same direction before an atom can exhibit a electromagnetic field. Furthermore, external electrical force (66/67) can alter the electromagnetic properties of a atom since electromagnetic force is dependent on the movement of charged particles in a electrostatic field. Voltage Intensifier circuit (190) of figure (3-23), now, allows voltage to dissociates water molecule (85) by overcoming electrostatic bonding force (qq') between unlike atoms (76/77) while restricting amp flow, as illustrated in (160) of Figure (3-26).

Electrical Polarization process

Placement of a pulse voltage potential (65) across Excitor plates (E1/E2) (voltage zones 66/67) of Figure (3-29) as to Figure (3-26) while inhibiting and preventing electron flow within voltage intensifier circuit (190) of Figure (3-23) causes water molecule (210) of Figure (3-27) to separate into its component parts (released hydrogen and oxygen gases) by pulling away (utilizing opposite attraction forces SS' and RR') its charged water molecule atoms (76/77), as illustrated in (160) of Figure (3-26).

Stationary "positive" electrical voltage-field (66) (voltage plate E1) not only attracts negative charged oxygen atom (76) but also pulls away negative charged covalent electrons (84) from water molecule (210). At the same time stationary "negative" electrical voltage field (67) (voltage plate E2) attracts positive charged hydrogen atoms (77a/b). Once negative electrically charged oxygen atom (76) is dislodged from water molecule (85), covalent bonding (sharing electrons between atoms) ceases to exist, switching-off and disrupting electrical attraction force (qq') between unlike atoms (76/77), as further illustrated in (160) of Figure (3-26).

Opposite polarity electrical attraction force (SS') continues to cause negative charged oxygen atom (76) to migrate to positive voltage-plate (E1) (positive voltage zone 66); while, at the same time, opposite polarity electrical attraction force (RR') causes positive charged hydrogen atoms (77a/b) to migrate in the opposite direction to negative voltage-plate (E2) (negative voltage zone 67) as step-charging voltage-wave (65) increases in voltage amplitude from several millivolts to several hundred volts during each pulse train (65a xxx 65n) which, in application, causes water molecule (210) of Figure (3-27) charged atoms (76/77) to elongate (increasing distance between unlike atoms 76/77) to the point where covalent hydrogen electrons (84) of Figure (3-27) breaks away from electrostatic force (qq'). Repetitive duplication of voltage pulse (65a xxx 65n) continues to separate or split apart other water molecules (85a xxx 85n) which, in turns, forms hydrogen (86) and oxygen (87) gas-mixture (88) of Figure (3-24). Dissociation of water molecule (85) by way of voltage stimulation (65) is herein called "The Electrical Polarization Process", as illustrated in (160) of Figure (3-26).

Resonant Action

Subjecting and exposing water molecule (85) to even higher voltage levels (xxx Vn) (up to and beyond several thousand volts) causes water bath (91) of Figure (3-30) as to Figure (3-25) to go into a state of ionization by allowing opposite polarity forces (TT') and (UU') to eject one or

RE: WFC Hydrogen Gas Management System   Memo WFC 422 DA

more electrons (92a xxx 92n) from water bath atoms (93). Intensified electrical attraction force (TT') causes dislodged negative charged electrons (92) to migrate to positive voltage-plate (E1) while electrical attraction force (UU') causes positive charged atom nucleus (94) to travel toward negative voltage-plate (E2). Applied electrical attraction force (TT') and (UU') always being of equal voltage intensity but opposite in electrical polarity as voltage amplitude (65) is attenuated.

Replication of higher voltage forces (TT') and (UU') during pulsing operations causes a continued release of other electrons (92a xxx 92n) from other water bath atoms (93a xxx 93n) which, in practice, increases electrical charges of water bath (91) since water bath (91) is a dielectric liquid. Water bath atoms (93a xxx 93n) having missing electrons (92) take-on a positive electrical charged (95) which is subject to and moved by negative electrical force (UU'); whereby, the liberated and free floating negative charge electrons (92) are subject to and move by positive electrical force (TT'). Applied together, electrical forces (TT') and (UU'), now, causes these moving electrically charged particles to superimpose a physical impact unto electrical polarization process (160), as shown in (170) of Figure (3-25)...thereby, increasing gas-yield (88) still further.

By attenuating voltage amplitude (Vo xxx Vn) in conjunction with pulse-width (65a xxx 65n) allows voltage intensifier circuit (190) of Figure (3-23) to tune-in and match the resonant characteristics or resonant frequency of water bath (91) since water bath (91) always maintains its dielectric properties during pulsing operations. At resonance, electrical polarization process (160) interacts uniformly with liberated charged particles (92/95) of Figure (3-25) to obtain a even higher gas-yield (88) at maximum voltage deflection (xxx Vn). The established resonant frequency is most generally in the audio range from 1 Khz up to and beyond 10 Khz and is dependent upon the amount of contaminants in natural water. Oscillating and superimposing electrical charged particles unto the Electrical Polarization process at a given pulse-frequency is, now, herein called "Resonant Action", as illustrated in (240) of Figure (3-25).

To reach maximum gas-yield (88) resonant cavity (170) of Figure (3-25) is shaped into a tubular structure (typically 0.50 inch diameter tube inserted into 0.75 inch diameter tube having a .0625 concentric air-gap 3 inches long) which functions as a longitudinal wave-guide to enhance particle movement in a lateral or angular displacement to applied voltage fields (66/67). Insulated

Stanley A. Meyer

RE: WFC Hydrogen Gas Management System   Memo WFC 422 DA

housing (72) prevents voltage coupling to water bath (68) which allows applied voltage amplitude (xxx Vn) to remain constant across water molecules (85a xxx 85n)...stabilizing gas production during voltage stimulation (65), as shown in (120) of Figure (3-24). To further prevent voltage fluctuation during resonant action, Phase Lock Loop technique of Pulse Indicator circuit (110) is utilized during pulsing operations. The resultant fuel-gas (88) is, now, transferred through Quenching Tube (96) of Figure (3-41) to, through and beyond Fuel Injectors (36) of Figure (3-1) for Hydrogen gas utilization.

In cases where applied voltage amplitude is to remain constant while promoting Resonant Action during control-state, incoming pulse train (64a xxx 64n) is varied independent of voltage amplitude to attenuate voltage intensity (66/67) which, in turns, effect gas production. In other applications, Voltage amplitude (66/67) in direct relationship to pulse-train (64a xxx 64n) may be varied together in a progressive manner to further control gas production. Or pulse-train (64a xxx 64n) can remain constant while voltage amplitude is varied. In all cases, Resonant Action is being promoted to product hydrogen gas on demand.

In terms of Longevity, voltage zones (E1/E2) are composed of or made of stainless steel T304 material which is chemically inert to hydrogen, oxygen, and ambient air gases (dissolved gases in water) being liberated from water bath (68) during voltage stimulation (65). Under actual certified laboratory testing stainless steel T304 life expectance (material decomposition) is .0001 per year since voltage (65) is a physical force, setting up a non-chemical environment since amps consumption is being restricted to a minimum and "no" electrolyte is added to water bath (68). In practice, stainless steel voltage plates (E1/E2) physically forms voltage zones (66/67) regardless of geometric shape or configuration of resonant cavity (170).

Under normal gas ignition or gas combustion process, released Fuel-Gases (88) of Figure (3-39) as to Figure (3-24) nets a thermal explosive energy yield (gtnt) of approximately 2 1/2 times greater than gasoline.

Stanley A. Meyer

RE: WFC Hydrogen Gas Management System					Memo WFC 422 DA

Gas Modulator Process

Dissolved air gases (97) of Figure (3-39) being uniformly released from water bath (85) via the Electrical Polarization Process (160) of Figure (3-26) is automatically intermixed with released hydrogen (86) and oxygen (87) gas atoms (also derived from water bath 85) to form Fuel-Gas mixture (88) of Figure (3-24) having a hydrogen gas burn-rate of approximately 47 centimeters per seconds (cm/sec) in ambient air, as illustrated in (330) of Figure (3-37). Volatility of hydrogen fuel-mixture or Fuel-Gas (88) is reduced from 325 cm/sec. to approximately 47 cm/sec. since ambient air gases (97) (dissolved air gases in water) is primarily composed of non-combustible gases (74) (such as nitrogen, argon, and other non-burnable gases) of Figure (3-39) which acts and performs as a "Gas Modulator" during thermal gas ignition (98), as illustrated in (320) of Figure (3-36). The non-combustible gases (74) physically retards and slows down the speed by which oxygen atom (87) unites with (covalent link up) hydrogen atoms (86a / 86b) to bring on and support gas ignition process (gas combustion process) (98), as further illustrated in (340) of Figure (3-38).

Water bath (68) of Figure (3-39) as to Figure (3-24), now, becomes and functions as a "Gas Mixing Regulator" since the highest possible thermal explosive energy yield (gtnt) obtainable from hydrogen during "normal" gas ignition (98) is the exact composition of water where two hydrogen atoms (86a / 86b) unite with oxygen atom (87).

Inherently, the utilization of the Electrical Polarization Process (160) of Figure (3-26) in conjunction with the use of chemically inert stainless steel (T304 material) voltage zones (E1 / E2) submerged in natural water (68) sustains and maintains gas mixing ratio (88) by simply preventing the consumption of both the hydrogen (86) and oxygen (87) gases by way of not encouraging "electrical heat" or "chemical interaction" associated with amp consumption. Remember, Electrical Polarization Process (160) is a physical process which uses opposite electrical polarity attraction force (qq') to perform work by disrupting and switching off the covalent bond between the unlike charged water molecule atoms.

To further reduce hydrogen burn-rate (330) of Figure (3-37) to other fossil-fuel burning levels, additional non-combustible gases (99a xxx 99n) (supplied via ambient air 101) is added to

Stanley A. Meyer

RE: WFC Hydrogen Gas Management System   Memo WFC 422 DA

gas-mixture (88) by way of gas ignition process (98) occurring inside internal combustion engine (55) piston cylinder (102), as illustrated in (340) of Figure (3-38). As fuel-gas (88) enters into engine cylinder (102) and is exposed to thermal gas ignition process (98), the incoming and moving fuel-gases (88) are converted into non-combustible gases (99) (gases passing through the gas combustion process) since both the hydrogen (86) and oxygen (87) gas atoms are being consumed during the formation of superheated water mist (103)...releasing thermal explosive energy (gtnt) which, in turns, causes piston-action to expel the newly formed non-combustible exhaust gases (99) for recycling.

The liberated and cooling exhaust gases (99) is, now, directed to hydrogen injector system (200) which systematically meter-mixes and superimposes a predetermined amount of non-burnable gases (99) of Figure (3-38) onto incoming ambient air gases (101) which is being directed to engine cylinder (102) to sustain and maintain both the "Gas Modulator Process" (320) of Figure (3-36) and the "Gas Ignition Process" (98), simultaneously. In essence,then, ambient air gases (101) becomes a endless supply of non-combustible gases (99A xxx 99n) during the gas ignition process.

The resultant and on-going Gas Modulator Process (320) of Figure (3-36),now, allows hydrogen fuel cell (120) of Figure (3-24) to be retrofitted to any conventional internal combustion engine (55) of Figure (3-1) without engine change by simply metering the proper amount of exhaust gases (99a xxx 99n) to comply with and co-equaling any type or different fossil-fuel burning levels, as further illustrated in (330) of Figure (3-37).

In terms of operability and performance, gas modulator process (320) continues to allow a conventional internal combustion engine (55) to run on ambient air gases; while, fuel-gas (88) not only cuts back and reduces oxygen extraction form ambient air (101) but produces a environmentally safe exhaust gases since non-combustible gases (99/74) from both ambient air gases (101) and Fuel-Gas (88) are thermally inert to gas ignition process (98).

RE: WFC Hydrogen Gas Management System   Memo WFC 422 DA

Gas Processor

To obtain higher energy-yields beyond the normal gas combustion process, ionized ambient air gases (104) of Figure (3-31) is, now, exposed to and intermixed with Fuel-Gases (88) prior to thermal gas ignition (98) of Figure (3-38), as illustrated in (240) of Figure (3-31). As ambient air gases (101) enters into and passes through air filter chamber (105) toward and beyond air gate assembly (GG), the moving air gases (101) are exposed to a high energy voltage fields (up to and beyond 2,000 volts) (106/107) of opposite electrical polarity which causes ambient air gases to become ionized gases (104), as illustrated in (260) of Figure (3-33). Positive electrical voltage field (106) causes negative charged orbital electrons (124a xxx) to be ejected from gas atom (101) due to opposite electrical attraction force (xx'); while, at the same time, negative electrical voltage field (107) exerts a second electrical attraction force (yy') on gas atom positive charged nucleus (108)...opposite electrical attraction forces (xx') and (yy') being of equal intensity, as further illustrated in (260) of Figure (3-33).

Once electron ejection occurs, the liberated and free floating electrons (117a xxx 117n) continues to migrate toward positive voltage zone (106); whereas, the newly formed ionized gas atom (having missing electrons) (104) continues to move onward and through air intake manifold (109) of Figure (3-31) to engine cylinder (102) of Figure (3-38).

The resultant ionized gas process (260) of Figure (3-33) is performed by Electron Extraction Circuit (270) of Figure (3-34) which function in like manner to Voltage Intensifier Circuit (60) of Figure (3-22) except amp consuming device (390) (such as a light bulb 112 placed between Resonant Charging Choke (56) and Gas Resonant Cavity (410) of Figure (3-34) is added to pulsing circuit (60) to cause and convert liberated electrons (117a xxx 117n) into radiant heat-energy (Kinetic energy) (113) in the form of light energy (114)...thereby preventing electrons (117a xxx 117n) from re-entering ionized gas process (260) ...destabilizing gas atom (101).

Repetitive formation of electrical voltage force or voltage intensity (65a xxx 65n) of Figure (3-21) attracts and causes liberated electrons (117a xxx 117n) to move electrically away from gas resonant cavity (410) and physically interact with light bulb filament (115) to initiate and perform

Stanley A. Meyer

kinetic conversion process (390), as further illustrated in (270) of Figure (3-34). The newly established and on-going electron conversion process (390) continues to aid ionized gas process (260) as other gas atoms (101a xxx 101n) are destabilized into ionized gas vapor (104a xxx 104n). The electron conversion process (390) is, of course, terminated when applied pulse voltage potential (65) is switched off. Pulsating voltage potential or voltage intensity (65a xx 65n) is adjusted, also, to "tune-in" to the resonant properties of ambient air gases (101) since ambient air gases (101) exhibits a dielectric value (air-gap of one inch resisting electron arc-over of up to 17,000 volts applied) between voltage plates (E3) and (E4), forming capacitor (410) of Figure (3-34).

Voltage fields (106/107) are physically configured (skin effect) by T304 stainless steel material to form voltage plates (E3/E4) of Figure (3-33) which are not only chemically inert to gas ionization process (260) but, also, forms tubular Gas Resonant Cavity (410) of Figure (3-34) having approximately the same size and shape of liquid resonant cavity (170) of Figure (3-25), as illustrated in (270) of Figure (3-34).

To further destabilize gas atom (104), emitted laser energy (electromagnetic energy having zero mass) (116) is, now, injected into Gas Resonant Cavity (410) via optical lens (121) and superimposed onto gas ionized process (260) and subsequently absorbed by gas atom nucleus (108), as illustrated in (260) of Figure (3-33) as to (270) of Figure (3-34). The absorbed laser energy (122) of Figure (3-35) not only causes ionized gas atom orbitals electrons (124) to be deflected away from gas atom nucleus (108) but, also, weakens electrostatic force (AA') between gas atom nucleus (108) and deflecting electrons (123a xxx) ...allowing even a greater number of electrons (117a xxx) to be ejected from ionized gas atom (104) being simultaneously subjected to Electron Extraction Process (260), as illustrated in (280) of Figure (3-35).

In essence, then, laser interaction (280) along with applied voltage process (260) causes gas atom (101) to go into sub-critical state (destabilizing the mass entity of a gas atom) since absorbed laser energy (122) prevents electrons re-capture (atoms accepting electrons) while interfacing circuit (270) dislodges, captures, and immediately consumes ejected electrons (117a xxx). In other words, ambient air gases (101) has, now, become a electromagnetically primed

RE: WFC Hydrogen Gas Management System Memo WFC 422 DA

destabilized gas atoms (104a xxx 104n) having missing electrons.

Solid state light-emitting diode (118) of Figure (3-33) arranged in a cluster-array (118a xxx 118n) mounted on printed circuit board (119) emits a discrete wave-length of light energy (electromagnetic energy) when light circuit assembly (420) of Figure (3-43) is electrically pulsed (126a xxx 126n) via variable pulsing circuit (125) in such a way as to vary light intensity (116) to match the light absorption rate of ionized gas (104), and, is determined with respect to the forward current through Led's (118) by

$$Rs = \frac{V_{in} - V_{led}}{I_{led}}$$ (Eq 15)

Where

I_{led}, is the specified forward current (typically 20ma per diode); V_{led} is the led voltage drop (typically 1.7 volts for red emitter's).

Ohm's law for led circuit in parallel array, and, is given by

$$P\,watts = Vcc \; It$$ (Eq 16)

Where

I_t is the forward current through led cluster-array; Vcc is volts applied (typically 5 volts)

Whereby

Laser or light intensity is variable as to duty cycle on/off pulse frequency from 1hz up to and beyond 10khz, and is given by

$$Le \sqrt{\frac{(ION)^2 \times T1}{T1 + T2}}$$ (Eq 17)

Le is light intensity in watts; T1 is current on-time; T2 is current off-time; and (ION) = RMS value of load current during on-period.

RE: WFC Hydrogen Gas Management System   Memo WFC 422 DA

In terms of assembly, gas resonant cavity (410), electron extraction circuit (270), optical lens (121) forms gas processor (260) of Figure (3-31). In retrospect to operational parameters, led's (118) light spectrum (extending from the visible into the Ultraviolet light region) can be selected for a given or predetermined electromagnetically gas priming application (280) since gas nucleus (108) is more responsive to coherent rather than diffused light source. Applied voltage amplitude (Va xxx Vn), applied voltage pulse frequency (65a xxx 65n), and applied current pulse train (126a xxx 126n) are design variable to "tune-in" to the resonant properties of gas atom (101) while stimulating and performing gas process (260) which attenuates electrical force (AA') of Figure (3-35) to disrupt the mass equilibrium of gas atom (104).

The resultant and newly formed sub-critical gas atoms (104a xxx 104n) are directed onward through air intake manifold (109) of Figure (3-31) to and beyond both exhaust gas metering port (370) and injector port (36) where metered fuel-gas (88), metered exhaust gases (99), and metered sub-critical gas atoms (104a xxx 104n) forms gas-mixture (103) entering engine cylinder (102), as illustrated in (240) of Figure (3-31) as to (340) of Figure (3-38).

Hydrogen Fracturing Process

Incoming processed hydrogen fuel gas (103) is, now, exposed to thermal spark ignition process (98) which triggers thermal explosive energy-yield (gtnt) (127) that causes piston-action (105) of Figure (3-38) to exceed normal gas combustion process associated with hydrogen to air mixture of gases in stable state. Thermal atomic interaction (127) is caused when sub-critical gas ions (104a xxx 104n) (derived from both water bath 68 and ambient air gases 101 fails to unite with or covalently link up or covalent bond with highly energized (laser primed) hydrogen atom (128).Sub-critical Oxygen atom (129) having less than the normal amount of covalent electrons (orbital electrons) is unable to reach "stable-state" (six to eight covalent electrons required) when the two hydrogen atoms (128 a/b) seek to form the water molecule during thermal gas ignition.

Absorbed laser energy (131) of hydrogen gas atom nucleus (133) weakens "electrical bonding" force (CC') between hydrogen atom electron (132) and hydrogen atom nucleus (133); while, at the same time, absorbed laser energy (135) prevents oxygen atom (129) from reaching "stable state" when electrical attraction force (BB') (opposite electrical attraction force being equivalent to the number of missing electrons) locks onto and pulls away hydrogen atom electron (132) while repelling force (DD') keeps the two positive charged nucleuses (133/136) apart. These "abnormal" and "unstable" conditions coupled with thermal interaction (gas ignition) under gas compression (137) of Figure (3-42) as to Figure (3-38) (fuel-gas 88 being compressed via piston-action 105) causes combustible gas atoms (129 and 128a/b) to decay...releasing thermal explosive energy (gtnt) (127) under control means. This atomic thermal-interaction between sub-critical combustible gas atoms (127 and 128a/b) is, now, herein after called "The Hydrogen fracturing Process."

Laser Distributor

Laser Distributor assembly (430) of Figure (3-44) functions in similar manner as Laser Accelerator (20) of Figure (3-10) except light-gate (141) of Figure (3-44) rotates in the same direction of Spark-rotor (142) and being displaced opposite to rotor blade (142), allowing intermixed process ambient air gases (101) and Fuel-Gases (88) to enter engine cylinder (102) of Figure (3-38), as illustrated in Injector Control Circuit (300) of Figure (3-4). Rotating light-gate triggering circuit assembly (430) sequentially activates Pulse Shaping Generator (440) of Figure (3-4) to produce a constant 50% Duty-cycle Pulse-Train (see Figure 3-16 once again) to Analog Voltage Generator (40) of Hydrogen Gas Management System (200) of Figure (3-1) as to Figure (3-5). Interlocking Laser Accelerator output (JJ) with Laser Distributor output (HH) of Figure (3-1) causes Fuel-Injectors (36) to be "Tuned" with both Air Management System (350) of Figure (3-2) and Hydrogen Gas Control Circuit (100) of Figure (3-5) to maintain constant Fuel-mixing Ratio (290) of Figure (3-3) during engine performance. As Laser Accelerator (JJ) advances toward "Peak" engine performance, Fuel-Injectors (36) open gate-time (on-time) increases proportionally. Opposite or reverse movement of Laser Accelerator (JJ) decreases Injectors (36) on-time which, in turns, reduces engine speed.

Impurity Extraction Process

Suspended and dissolved water contaminates (144a xxx 144n) (typically 20 ppm to 40 ppm in natural water) of Figure (3-24) being uniformly released from and superimposed onto remaining water bath (68) during Resonant Action (170) are directed to and passes through water inlet line (145)...causing liberated, moving, and free-floating micro-sized contaminates (144a xxx 144n) to be deposited inside Electrostatic Filter Assembly (440) and subsequently exposed to opposite electrical voltage fields (148/152), as so illustrated in Figure (3-45).

Negative electrically charged contaminates (157a xxx 157n) migrates to and entrapped by positive electrical voltage post (147); while, simultaneously, positive electrical charged contaminates (158a xxx 158n) are attracted to and entrapped by negative electrical voltage post

RE: WFC Hydrogen Gas Management System							Memo WFC 422 DA

(151)...thereby, extracting contaminates (144a xxx 144n) from water bath (68)...producing purified water bath (156) which is recycled back into Fuel Cell (120) of Figure (3-24) since Resonant Action (170) also function as and performs as a water pump (Gas rising).

Exposing water contaminates (144a xxx 144n) to applied voltage fields (E1/E2) not only produces electrical charged contaminates (157/158); but, also, kills bacteria that might be present in water bath (68). Periodically back-flushing (rinsing out) is all that is require to ensure and sustain Electrostatic Filter Process (440).

Steam Resonator

To further ensure proper Fuel Cell operational performance during frigid or below freezing weather conditions, Steam Resonator assembly (450) of Figure (3-46) is inserted into Fuel Cell (120) of Figure (3-24) and thermostatically activated via Voltage Intensifier Circuit (165) which directly applies an alternate or opposite (166/167) electrical voltage pulses (during amp restriction) in a sequential manner across voltage plates E5/E6.

Once positive energized, water molecule (210) of water bath (68) is deflected toward voltage surface (E5) via both opposite electrical attraction force (162) and electrical repelling force (161). By simply reversing to a negative voltage pulse (167),now, causes water molecule (210) to be deflected in the opposite direction toward voltage surface (E6)...producing kinetic energy (165) (particle impact) which, in turns, heats water bath (68). Repetitive formation of opposite voltage pulses (166/167) at a given pulse-frequency continues to heat water bath (68) until a desired temperature is reach.

Diesel Application

By simply adjusting Fuel-gases (88) of figure (3-38) burn-rate (330) of Figure (3-37) from (43 ~ 37 cm/s) (Gasoline) to (40 ~ 35 cm/s) (Diesel) burning levels, now, allows WFC Hydrogen Gas Management System to be directly retrofitted to conventional Diesel Engines since the newly established modulated Hydrogen Fuel-mixture (88 ~ Diesel) co-equals spark-ignition ratio (typically .35 ~.39) of standard Diesel-Fuel (Fossil) under compression.

Aviation Application

Likewise, WFC Hydrogen Gas Management System is ideally suited as a retrofit energy

RE: WFC Hydrogen Gas Management System   Memo WFC 422 DA

system to both reciprocating and jet engines associated with the aviation industry... but in different ways: reciprocating WFC Fuel-kits can be similar to Car design (340); whereas, Water Fuel Injector kits (10) of figure (4-1) can alternately be used as a self-contained Fuel-unit having no pre-pressurized vessel which converts water directly in thermal explosive energy (gtnt) on demand, as illustrated in WFC memo 423 DA.

In terms of mechanical interfacing: Water Fuel Injector Assembly (10) of Figure (4-1) can replace standard fuel-injector ports of existing jet engines as shown in (150) of Figure (4-13); or be utilized in Furnace Nozzle Assembly (140) of Figure (4-12) for grain dryers or conventional heating systems; or be used to produce rocket-thrust, as illustrated in (160) of Figure (4-14); or be used as a spark plug injector nozzle (130) of Figure (4-11) for both gasoline and diesel engines...to mention a few.

Operational Parameters

Coupling and subsequently integrating Hydrogen Gas Management (GMS) system (440) with either Hydrogen Fuel-Gas Assembly (450) of Figure (3-1) or Water Fuel Injector Assembly (10) of Figure (4-1) as to Water Fuel Management (WFMS) System (40) of Figure (4-2), now, sets up a full engineering system-approach on how to use Water as a "New" fuel-source. It's design concept and system application complies with the "laws of economics" since micro-chip electronics and plastic mold injection technology help ensure performance reliability and usage ...especially since Fuel Cell (120) of Figure (3-24) is miniaturized to Water Fuel Injector Plug (40) of Figure (4-2), as further illustrated in WFC memo 423 DA.

Stanley A. Meyer

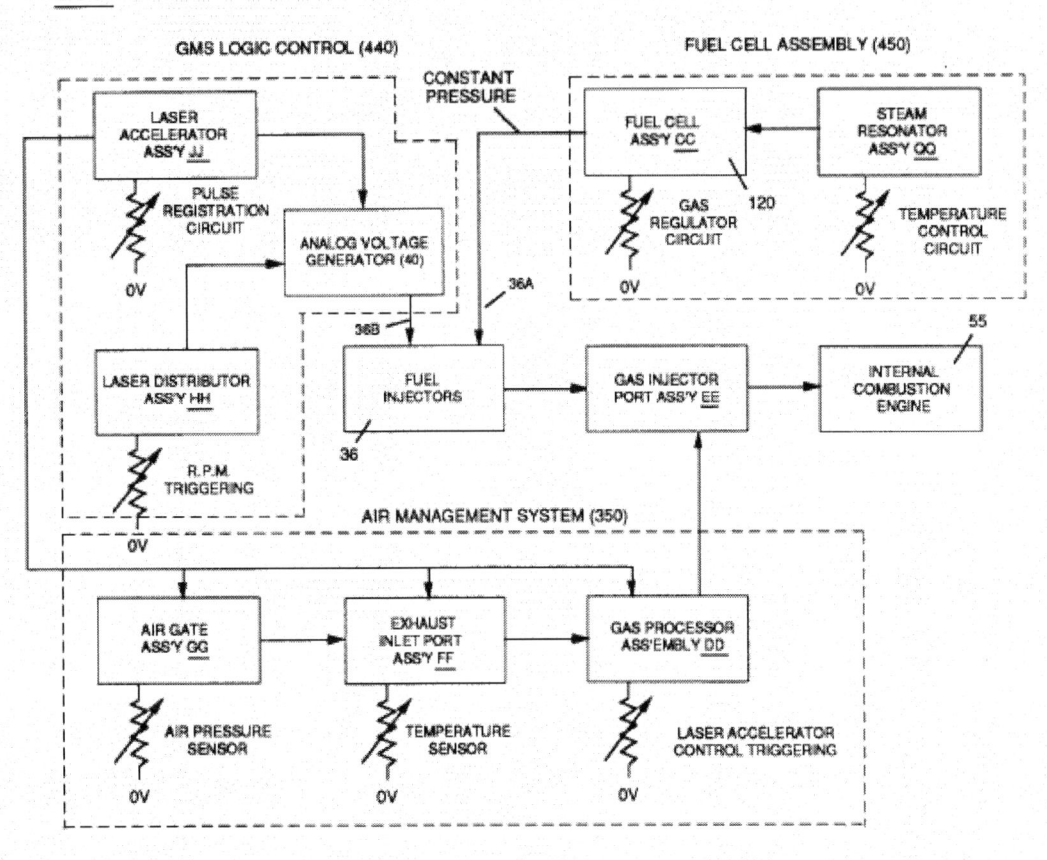

FIGURE 3-1: HYDROGEN GAS MANAGEMENT SYSTEM

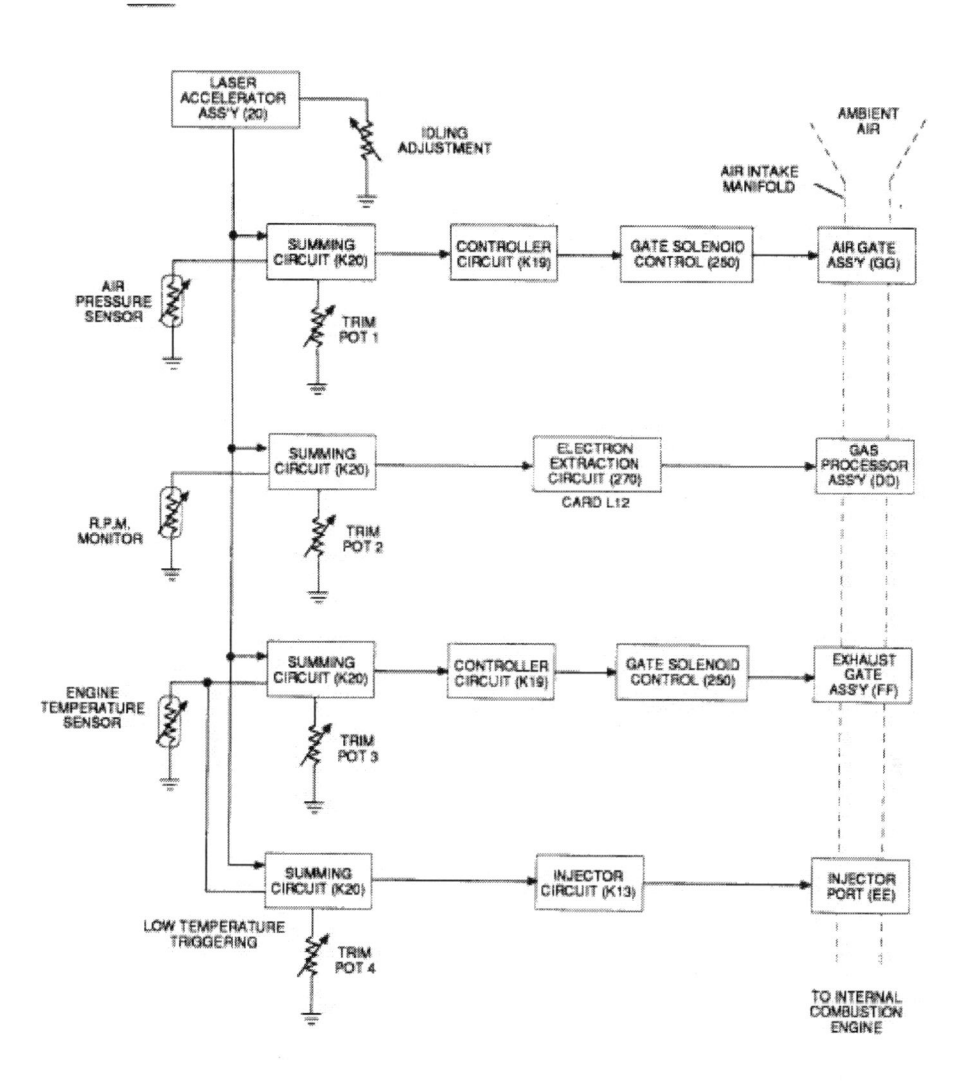

FIGURE 3-2: AIR MANAGEMENT SYSTEM

FIGURE 3-3: HYDROGEN GAS MANAGEMENT PERFORMANCE CURVE

(290) Adjusted Hydrogen Burn Rate 43 ~ 37 cm/s — Performance Curve (Linear Movement), from Idling Speed to Maximum R.P.M., across WFC Hydrogen Gas Management System.

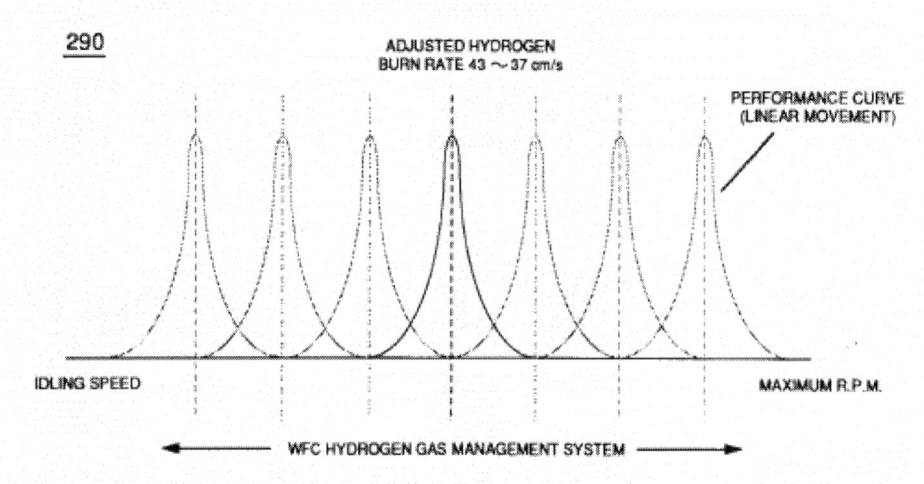

NOTE: HYDROGEN GAS BURN-RATE REMAINS CONSTANT REGARDLESS OF R.P.M.

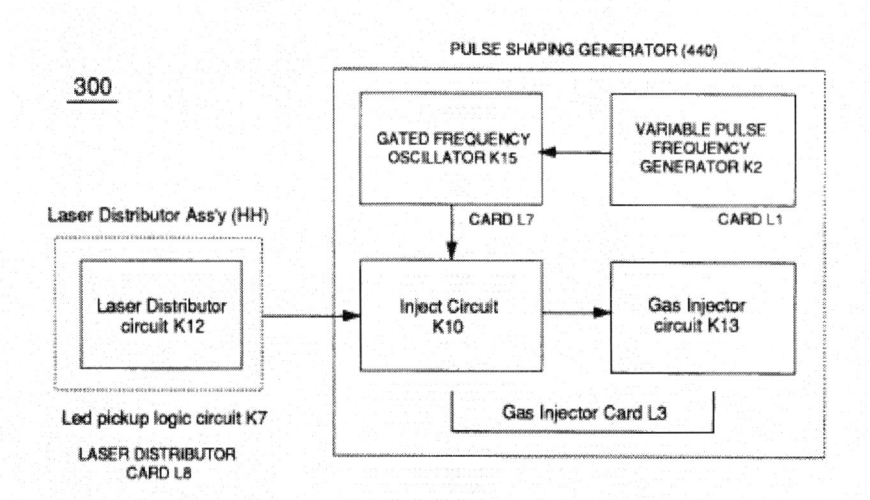

FIGURE 3-4: INJECTOR CONTROL CIRCUIT

(300) Pulse Shaping Generator (440): Gated Frequency Oscillator K15 (Card L7), Variable Pulse Frequency Generator K2 (Card L1), Inject Circuit K10, Gas Injector circuit K13, Gas Injector Card L3.

Laser Distributor Ass'y (HH): Laser Distributor circuit K12, Led pickup logic circuit K7, LASER DISTRIBUTOR CARD L8.

RE: WFC Hydrogen Gas Management System   Memo WFC 422 DA

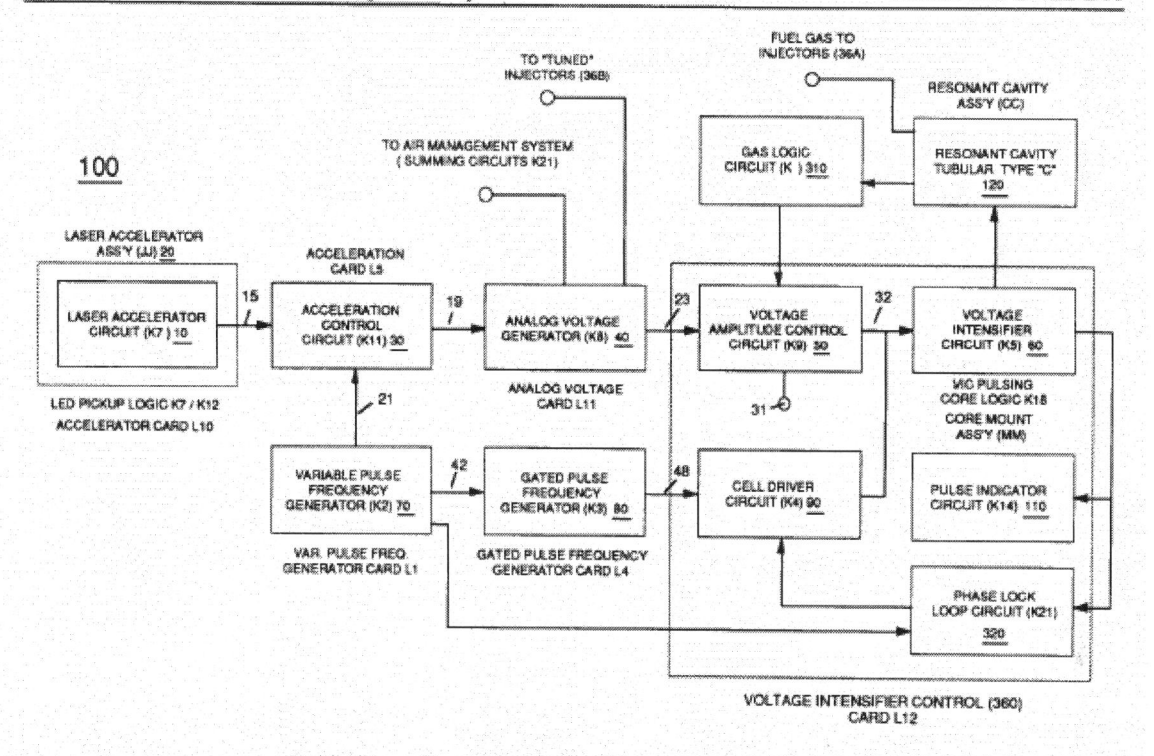

FIGURE 3-5: HYDROGEN GAS CONTROL CIRCUIT

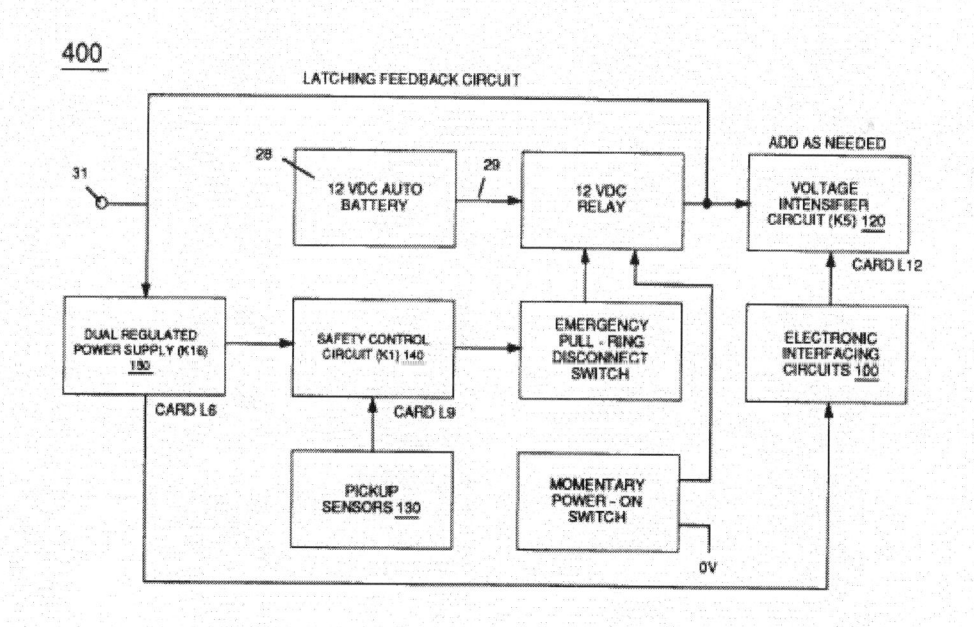

FIGURE 3-6: SAFETY INTERLOCK CIRCUIT

Stanley A. Meyer   3-30

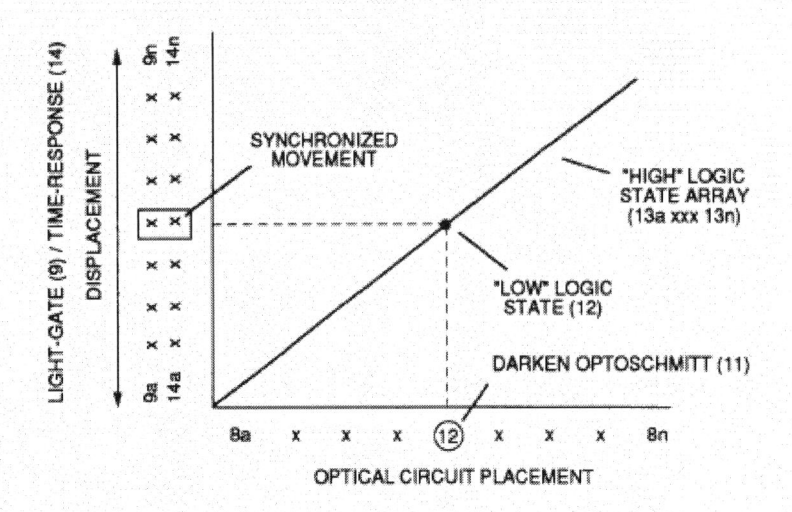

FIGURE 3-7: LASER ACCELERATOR CONTROL FUNCTION

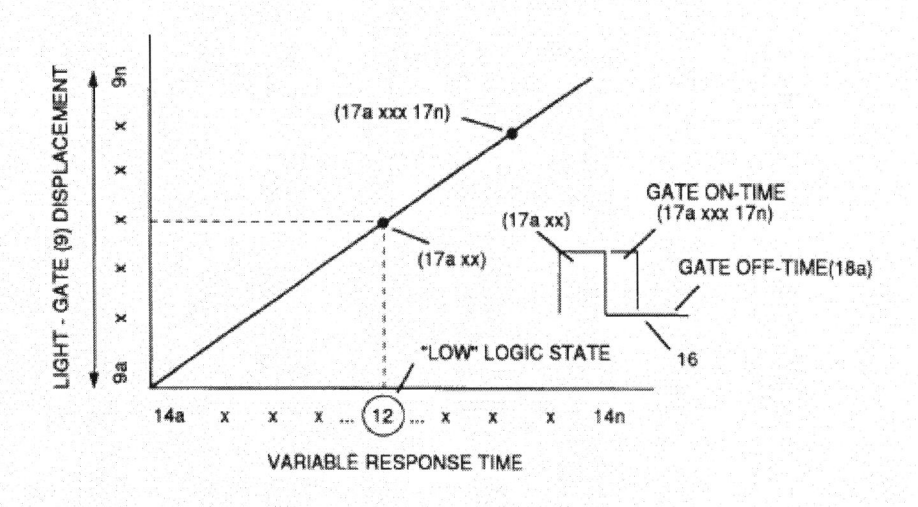

FIGURE 3-8: VARIABLE PULSE WIDTH

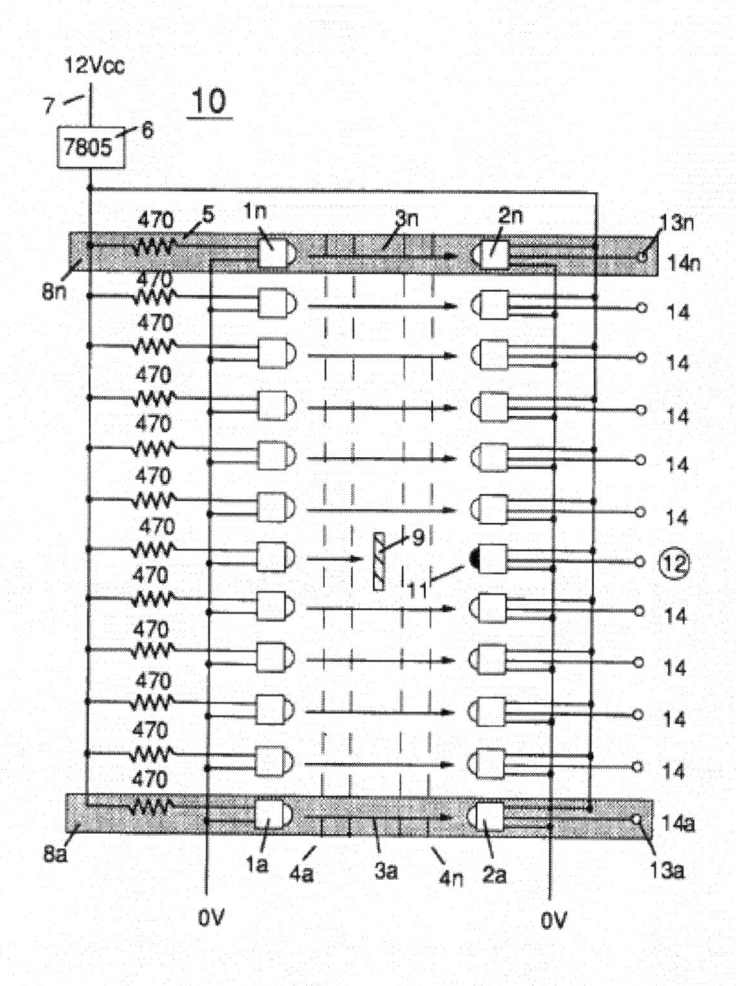

FIGURE 3-9: LED PICKUP CIRCUIT

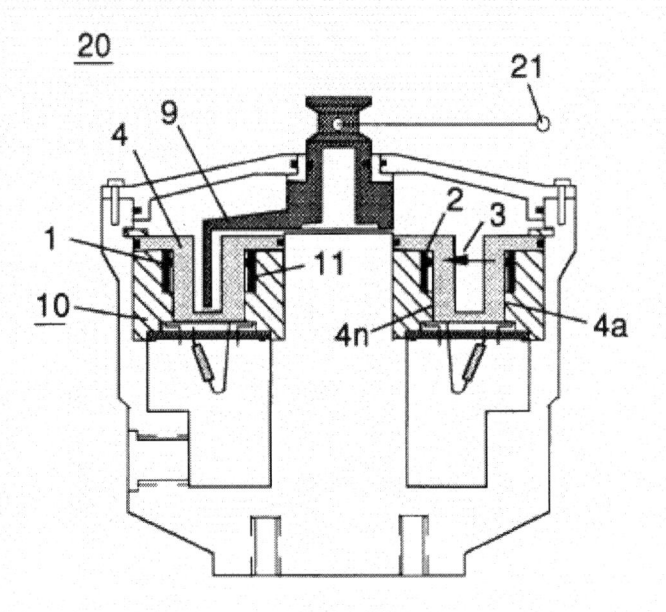

FIGURE 3-10: LASER ACCELERATOR

CONTINUOUS VARIABLE PULSE - TRAIN

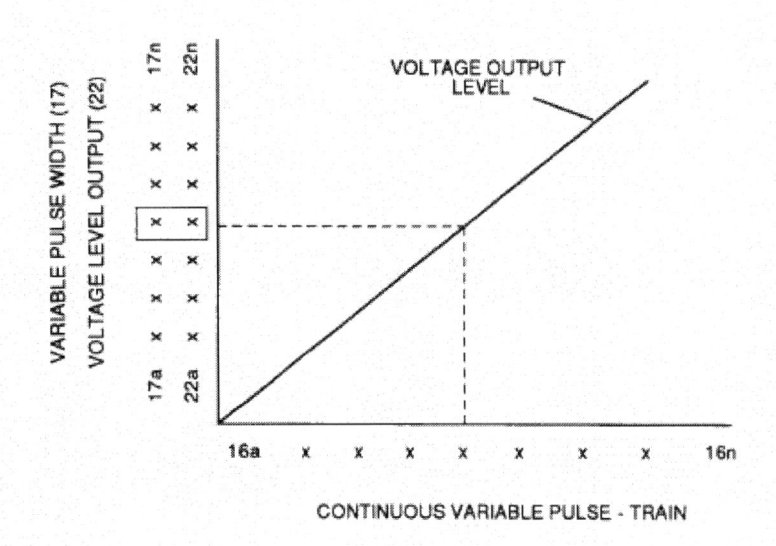

FIGURE 3-11: ANALOG VOLTAGE GENERATOR

PULSE LOGIC CIRCUIT (10)

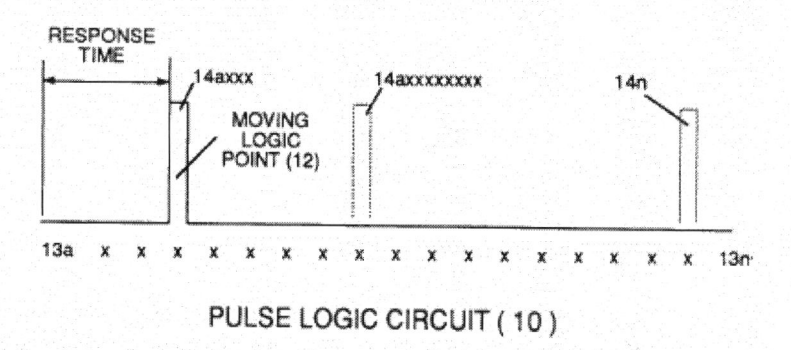

FIGURE 3-12: ACCELERATION CONTROL

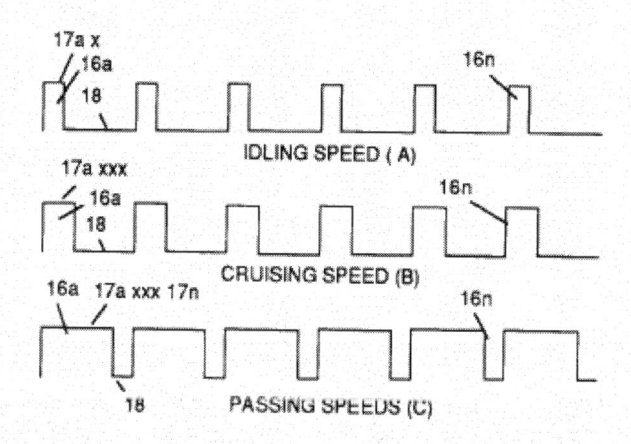

IDLING SPEED (A)

CRUISING SPEED (B)

PASSING SPEEDS (C)

FIGURE 3-13: SPEED CONTROL

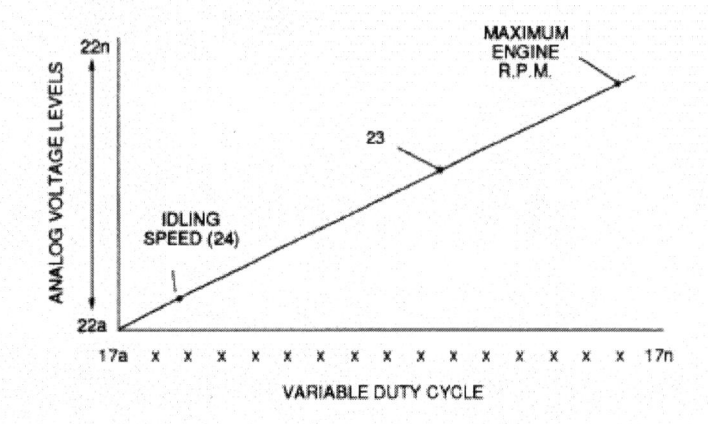

FIGURE 3-14: GAS VOLUME CONTROL

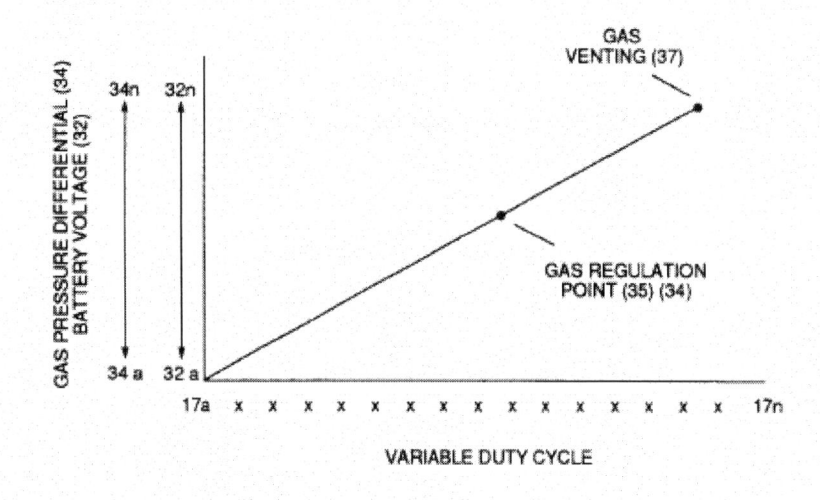

FIGURE 3-15: GAS REGULATOR CONTROL

RE: WFC Hydrogen Gas Management System   Memo WFC 422 DA

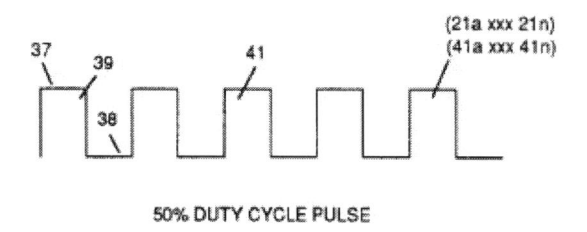

50% DUTY CYCLE PULSE

FIGURE 3-16: VARIABLE CLOCK PULSE TRAIN

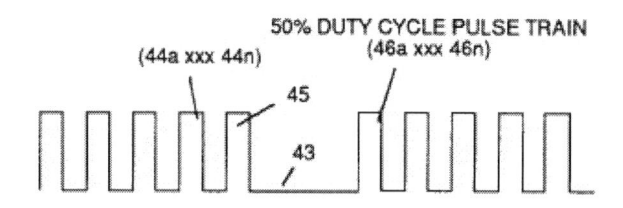

FIGURE 3-17: GATED PULSE TRAIN

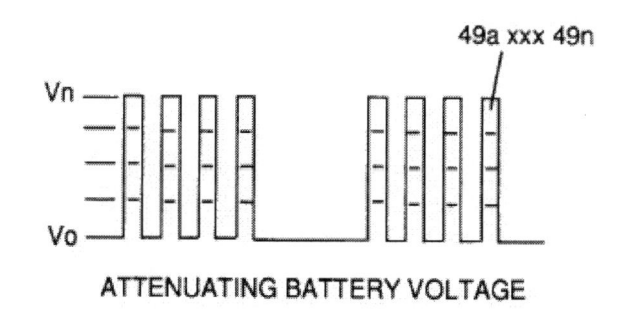

ATTENUATING BATTERY VOLTAGE

FIGURE 3-18: VARIABLE AMPLITUDE PULSE TRAIN

RE: WFC Hydrogen Gas Management System   Memo WFC 422 DA

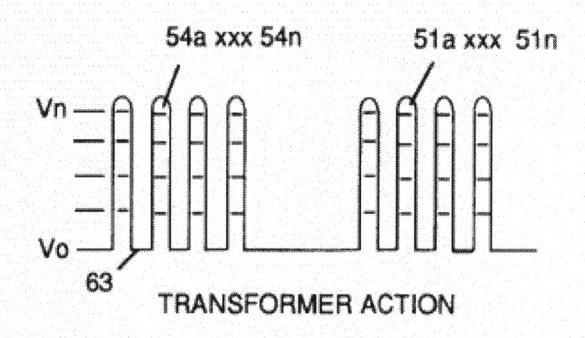

TRANSFORMER ACTION

FIGURE 3-19: SETUP VOLTAGE PULSE TRAIN

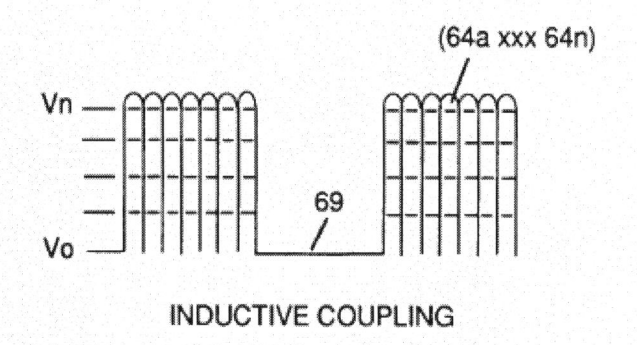

INDUCTIVE COUPLING

FIGURE 3-20: GATED UNIPOLAR PULSE TRAIN

ELECTRICAL STEP CHARGING EFFECT

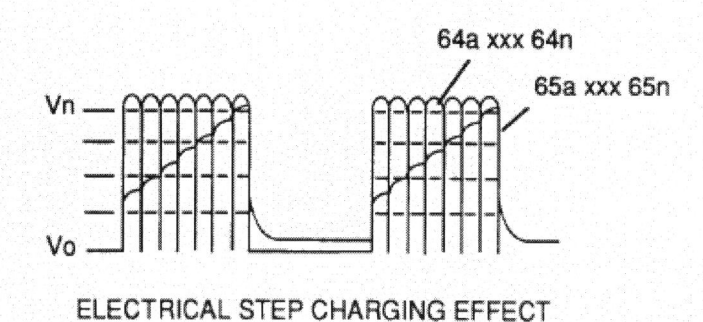

FIGURE 3-21: RESONANT CHARGING PULSE TRAIN

Stanley A. Meyer   3 - 37

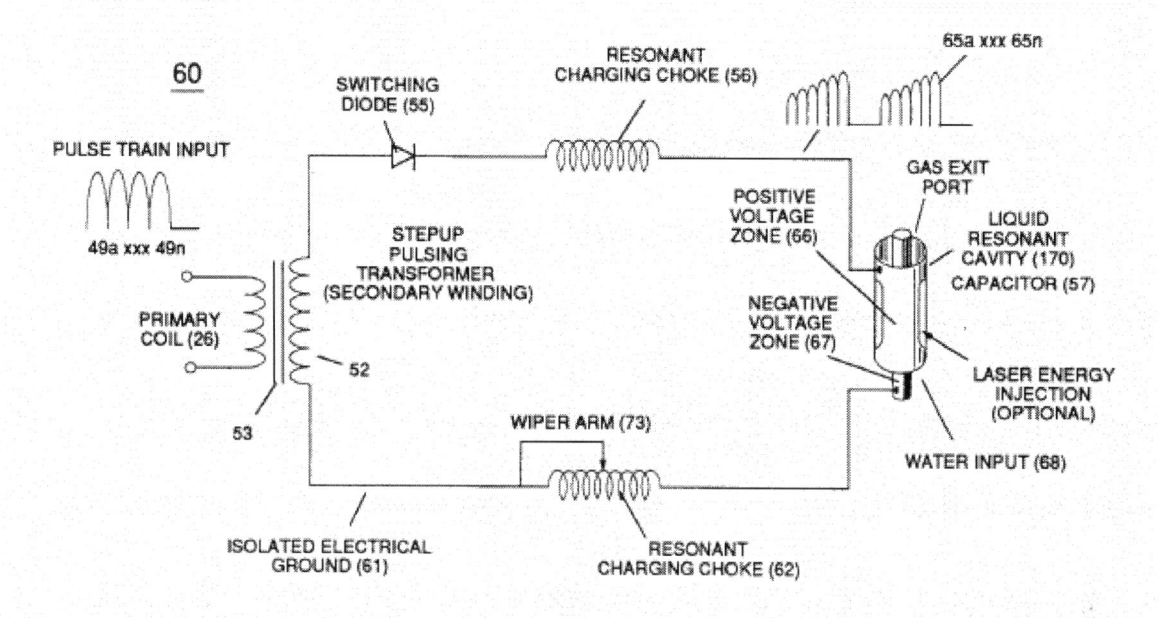

RE: WFC Hydrogen Gas Management System | Memo WFC 422 DA

FIGURE 3-22: VOLTAGE INTENSIFIER CIRCUIT

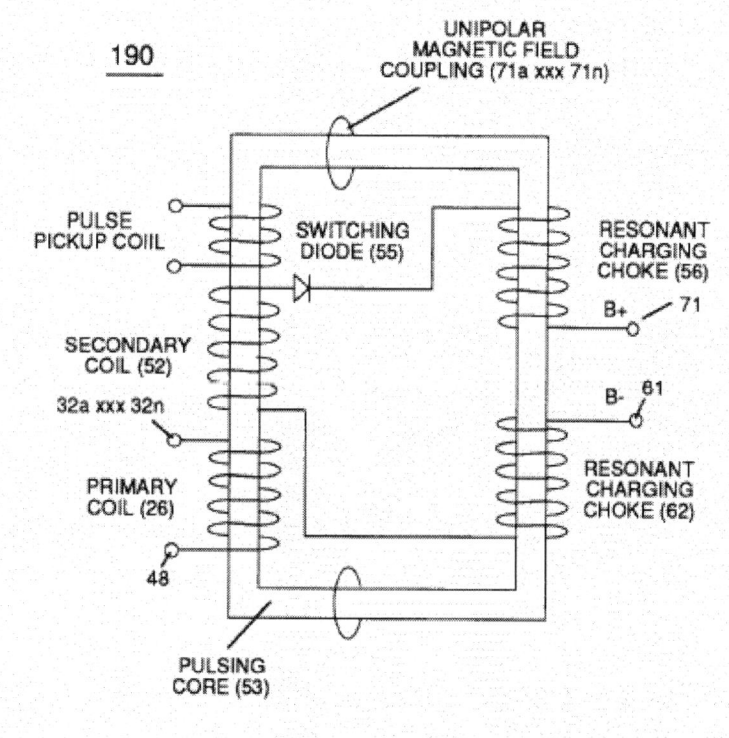

FIGURE 3-23 : PULSING CORE CONFIGURATION

Stanley A. Meyer | 3 - 38

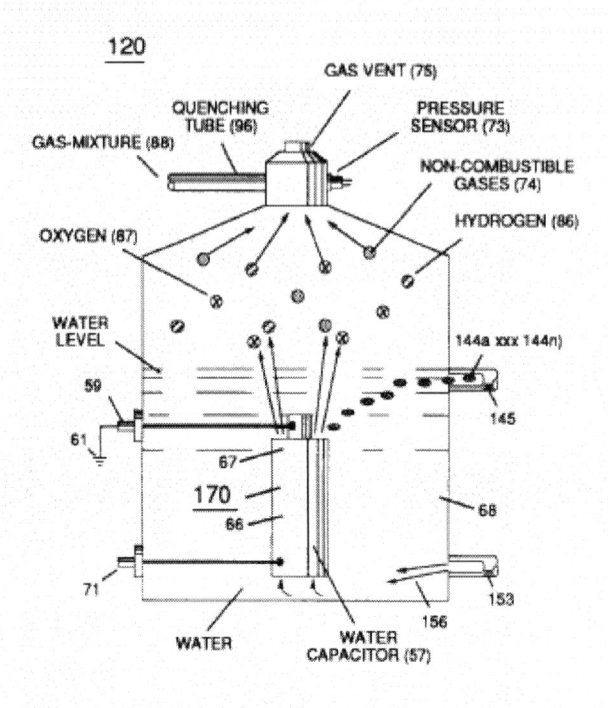

RE: WFC Hydrogen Gas Management System   Memo WFC 422 DA

FIGURE 3-24: FUEL CELL

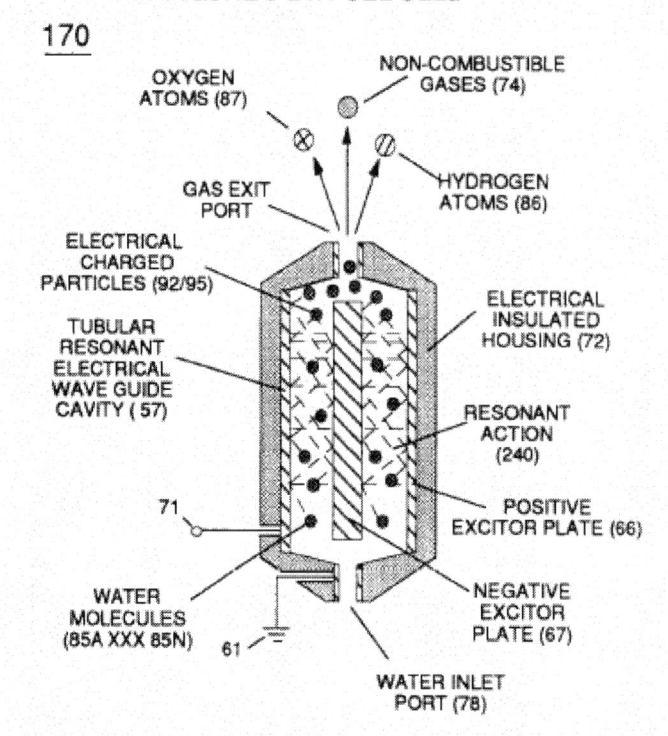

FIGURE 3-25: RESONANT CAVITY

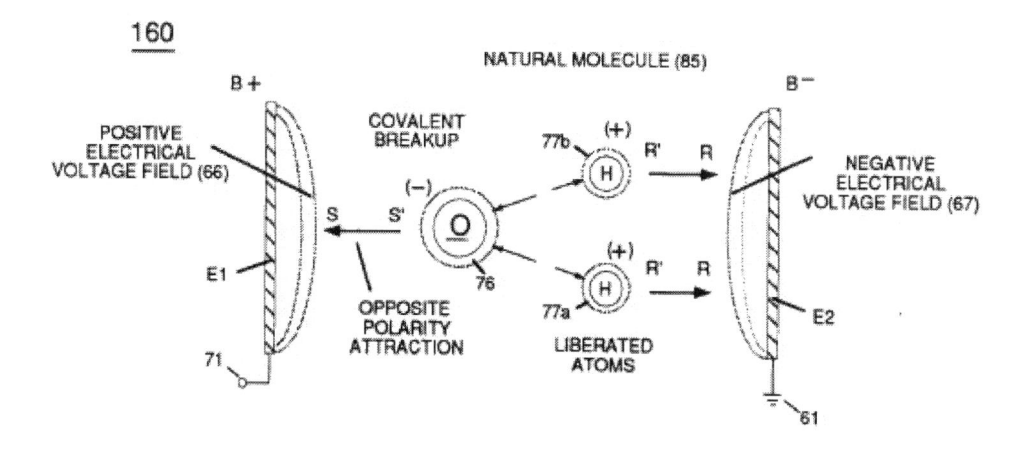

FIGURE 3-26: ELECTRICAL POLARIZATION PROCESS

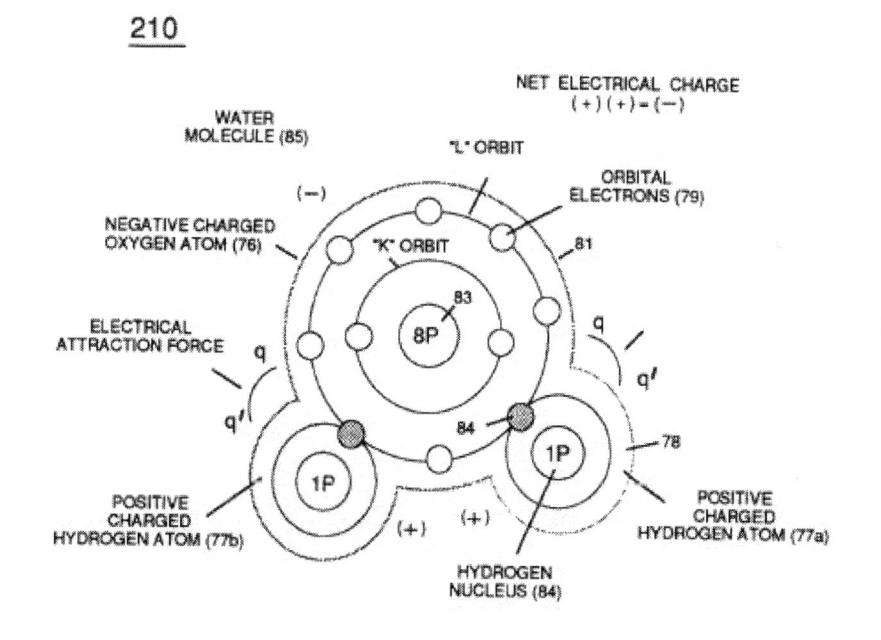

FIGURE 3-27: ELECTRICALLY CHARGED WATER MOLECULE

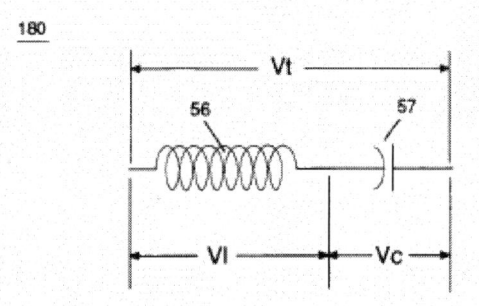

FIGURE 3-28: LC CIRCUIT SCHEMATIC

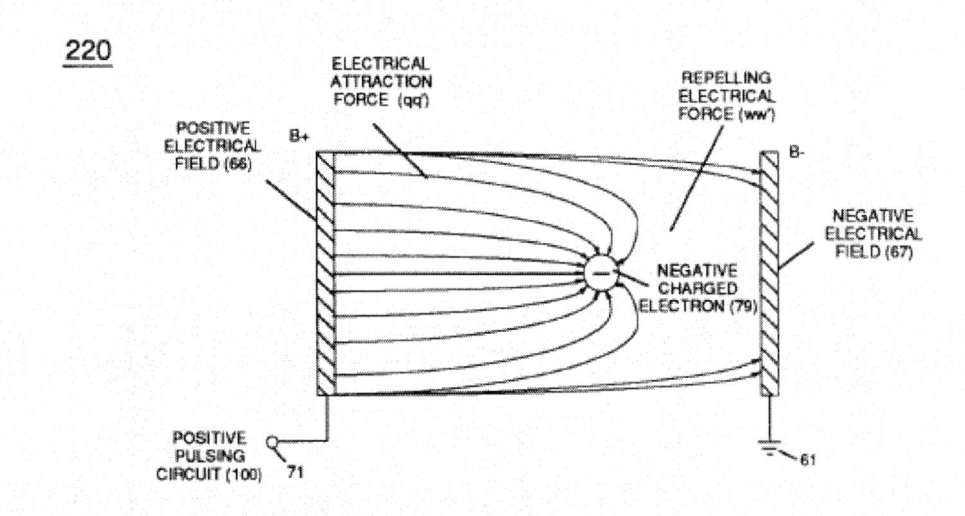

FIGURE 3-29: VOLTAGE DYNAMICS

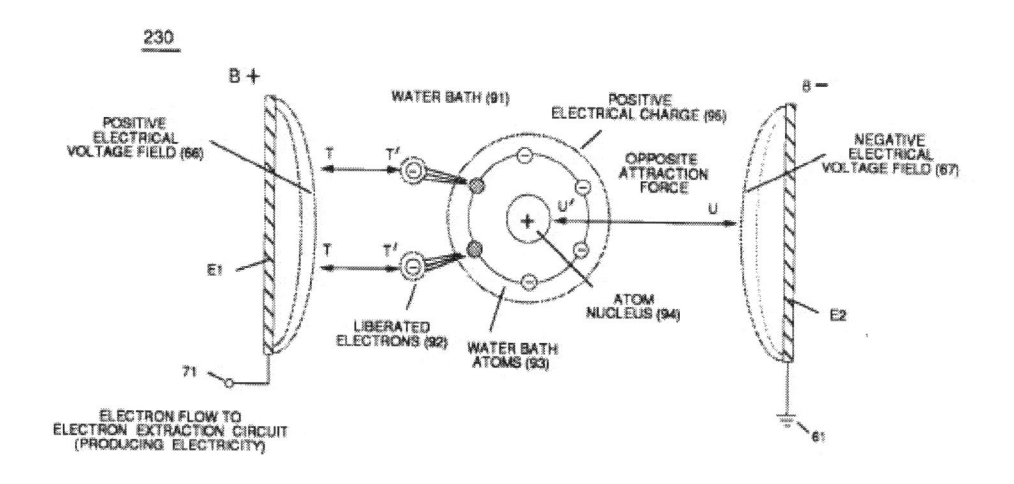

RE: WFC Hydrogen Gas Management System　　　　　　　　　　Memo WFC 422 DA

FIGURE 3-30: ELECTRON EJECTION

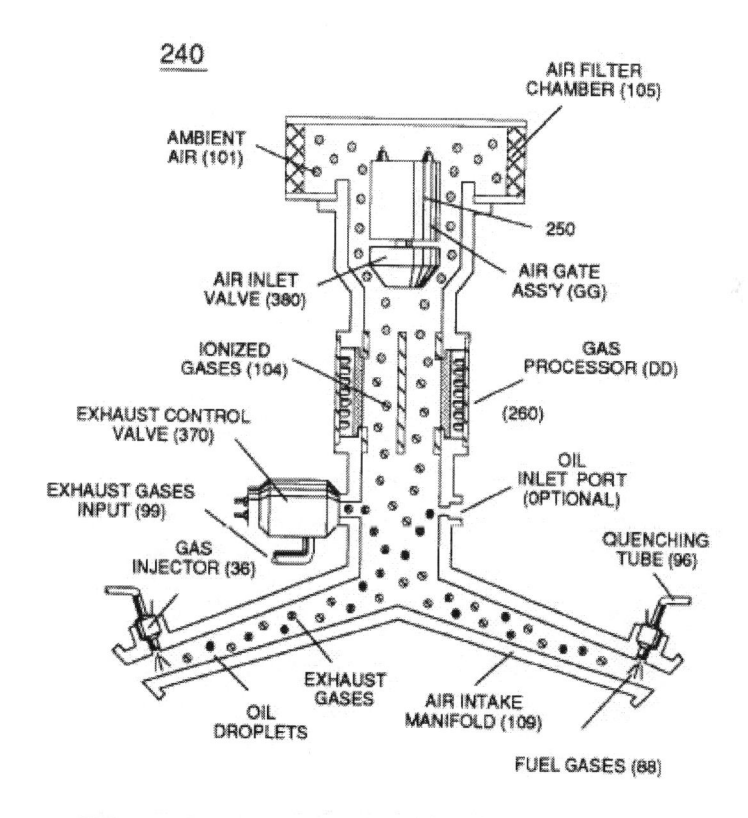

FIGURE 3-31: HYDROGEN GAS INJECTION SYSTEM

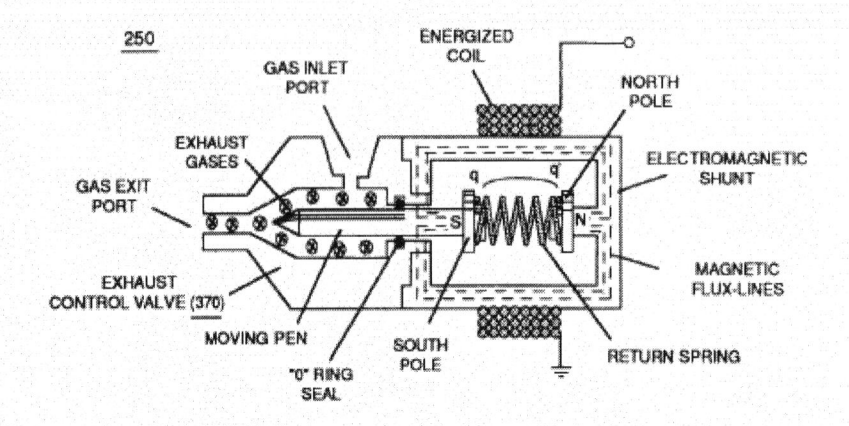

RE: WFC Hydrogen Gas Management System　　　　　　　　　Memo WFC 422 DA

FIGURE 3-32: EXHAUST INLET PORT

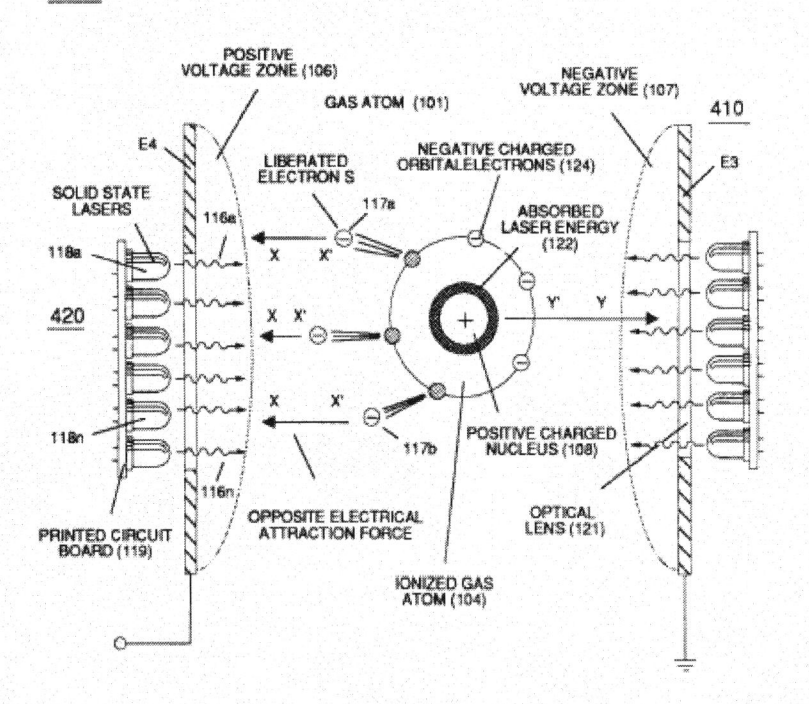

FIGURE 3-33: GAS PROCESSOR

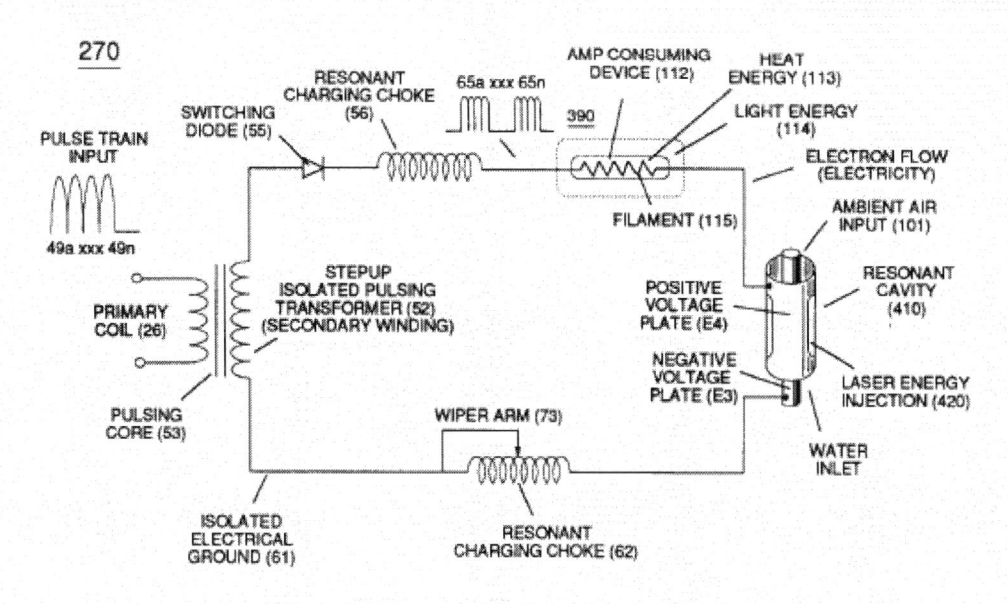

FIGURE 3-34: ELECTRON EXTRACTION CIRCUIT

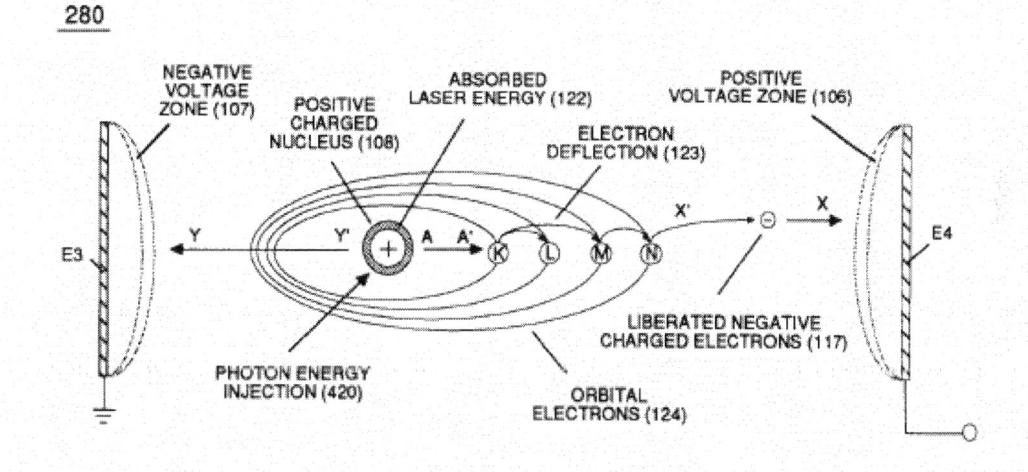

FIGURE 3-35: LASER INTERACTION

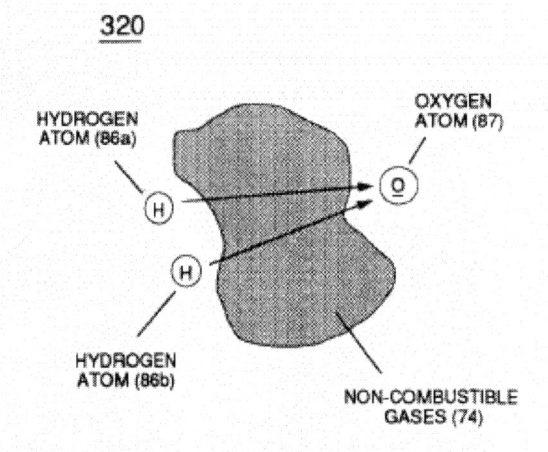

FIGURE 3-36: GAS MODULATOR

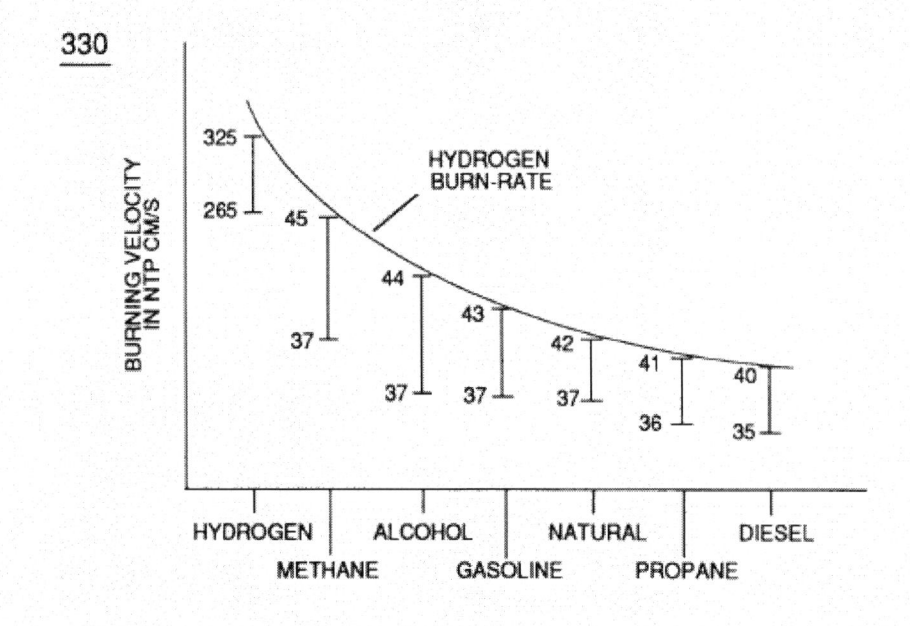

FIGURE 3-37: HYDROGEN BURN-RATE

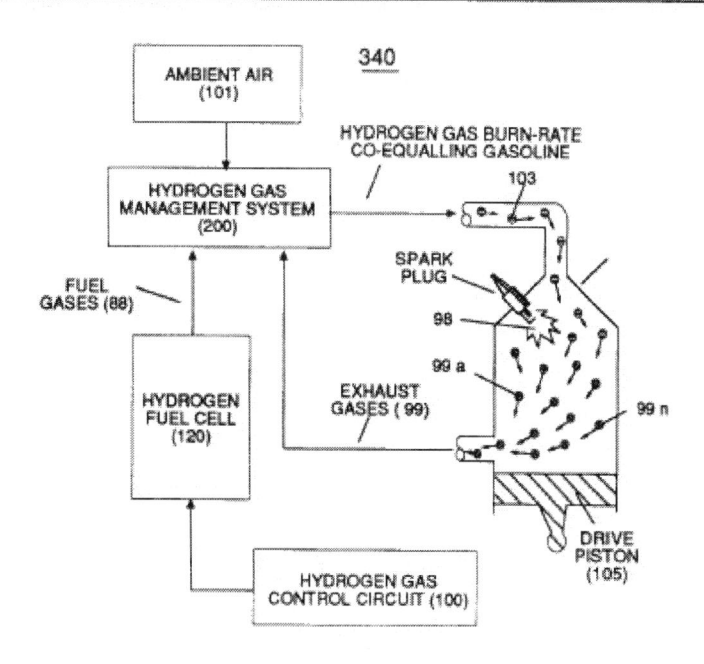

RE: WFC Hydrogen Gas Management System　　　　　　　　　Memo WFC 422 DA

FIGURE 3-38: RETROFIT ENERGY SYSTEM

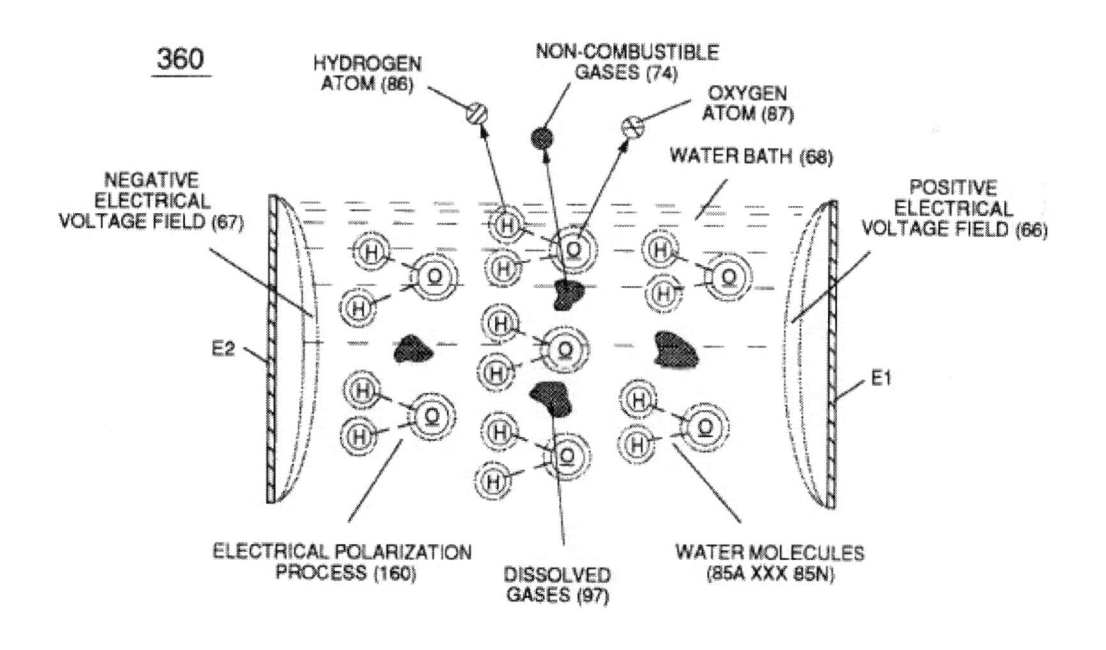

FIGURE 3-39: MULTI-GAS GENERATOR

370

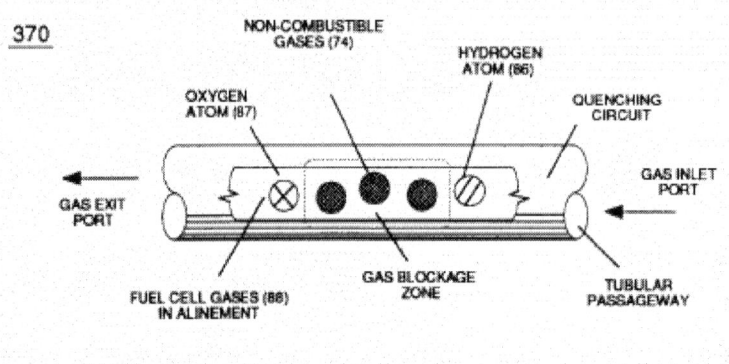

NOTE:
1) OXYGEN ATOM MUST UNITE WITH HYDROGEN ATOMS TO CAUSE GAS IGNITION.
2) TUBULAR PASSAGEWAY PREVENTS MOVING GAS ATOMS FROM REGROUPING.

FIGURE 3-40: QUENCHING ACTION

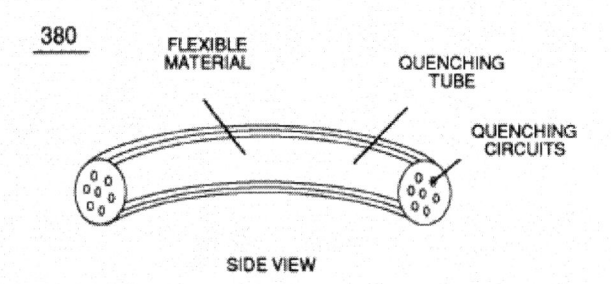

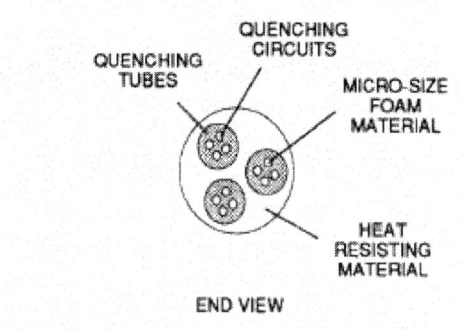

380

SIDE VIEW

END VIEW

FIGURE 3-41: SPARK ARRESTING GAS LINE

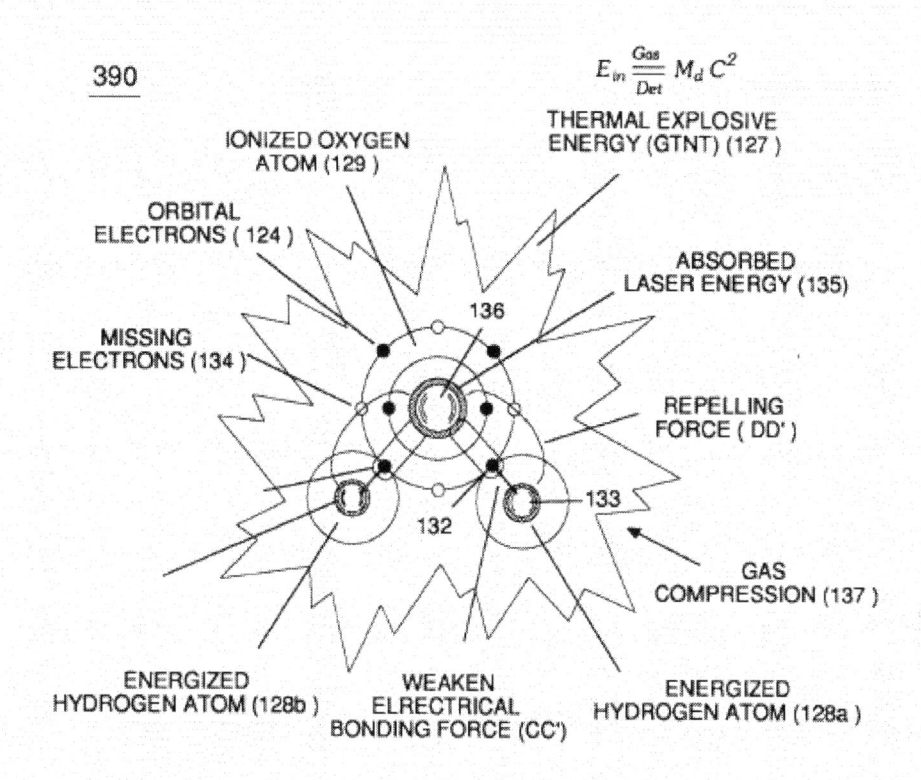

RE: WFC Hydrogen Gas Management System   Memo WFC 422 DA

390

$$E_{in} \underset{Det}{\overset{Gas}{=\!=\!=}} M_d\, C^2$$

- IONIZED OXYGEN ATOM (129)
- THERMAL EXPLOSIVE ENERGY (GTNT) (127)
- ORBITAL ELECTRONS (124)
- ABSORBED LASER ENERGY (135)
- 136
- MISSING ELECTRONS (134)
- REPELLING FORCE (DD')
- 133
- 132
- GAS COMPRESSION (137)
- ENERGIZED HYDROGEN ATOM (128b)
- WEAKEN ELRECTRICAL BONDING FORCE (CC')
- ENERGIZED HYDROGEN ATOM (128a)

FIGURE 3-42: HYDROGEN FRACTURING PROCESS

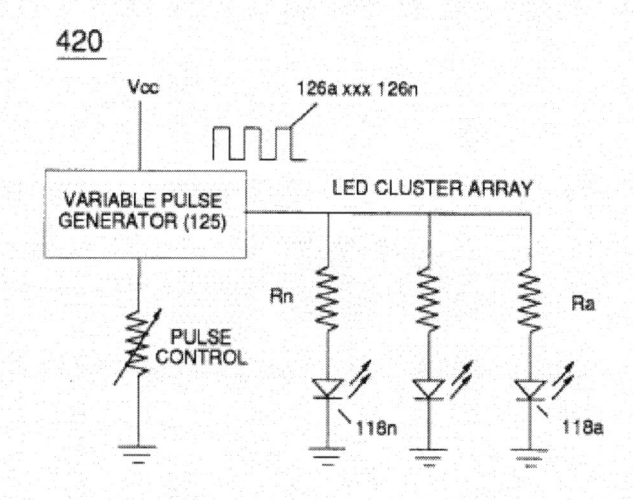

420

- Vcc
- 126a xxx 126n
- VARIABLE PULSE GENERATOR (125)
- LED CLUSTER ARRAY
- Rn
- Ra
- PULSE CONTROL
- 118n
- 118a

FIGURE 3-43: SOLID STATE LASER ASS'Y

Stanley A. Meyer   3 - 48

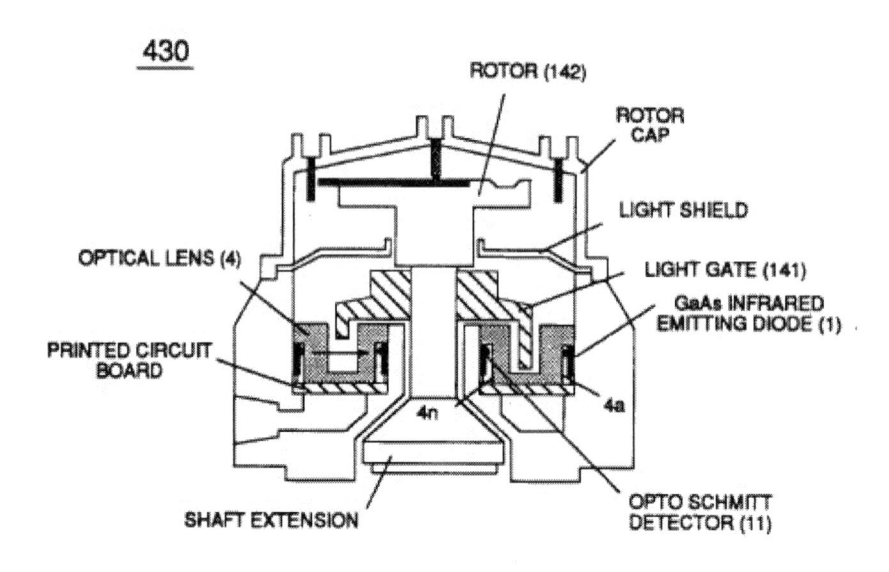

RE: WFC Hydrogen Gas Management System　　　　　　　　　　　Memo WFC 422 DA

FIGURE 3-44: LASER DISTRIBUTOR

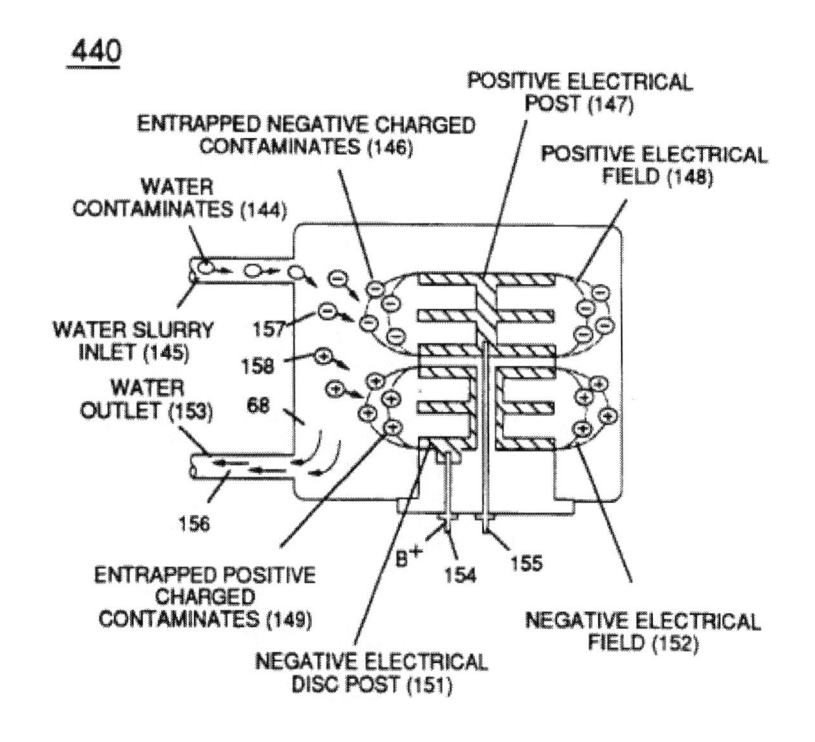

FIGURE 3-45: ELECTROSTATIC FILTER ASS'Y

RE: WFC Hydrogen Gas Management System

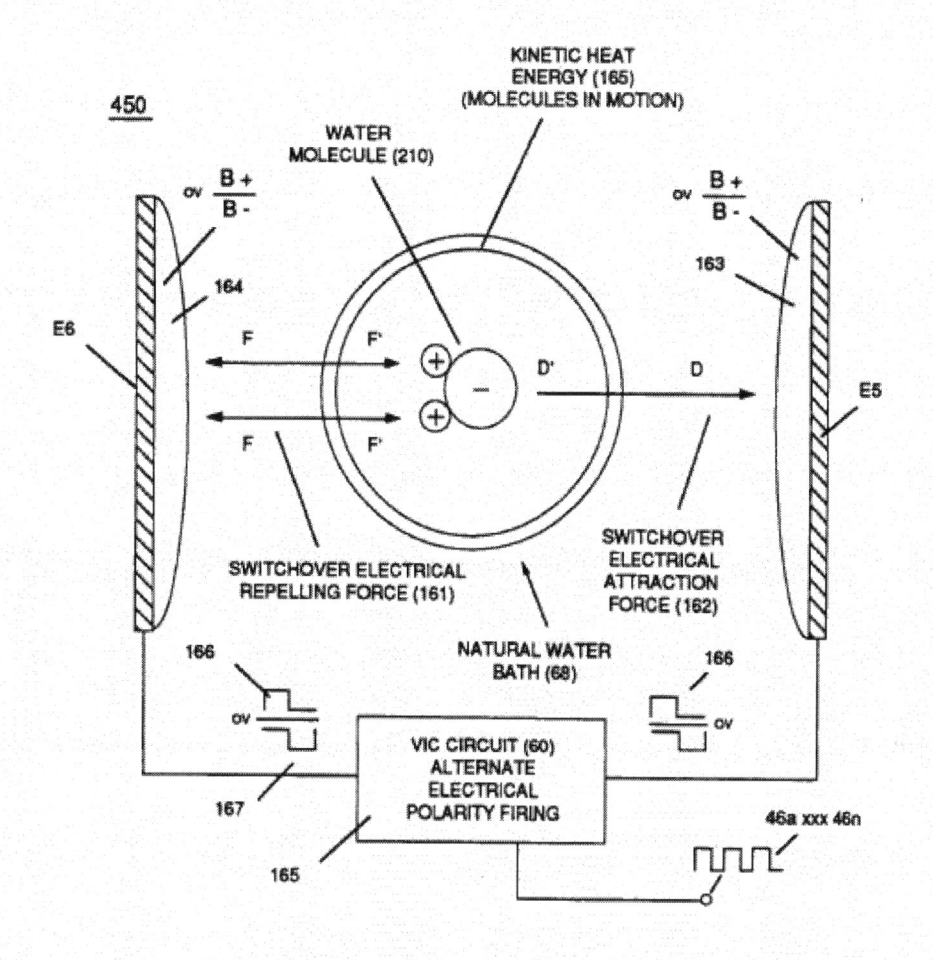

FIGURE 3-46: STEAM RESONATOR

WFC PROJECT 422 DA

- WATER INJECT GATE CONTROL
- GAS PROCESSOR
- RESONANT CAVITY FUEL CELL
- CONSTANT DISPLACEMENT WATER PUMP
- BUTTERFLY AIR GATE
- HYBRID LASER DISTRIBUTOR
- GAS INJECT GATE CONTROL
- WATER FUEL GAS INJECTORS ® (SPARK PLUG REPLACEMENT)
- DIFFERENTIAL SOLENOID GATE CONTROLS

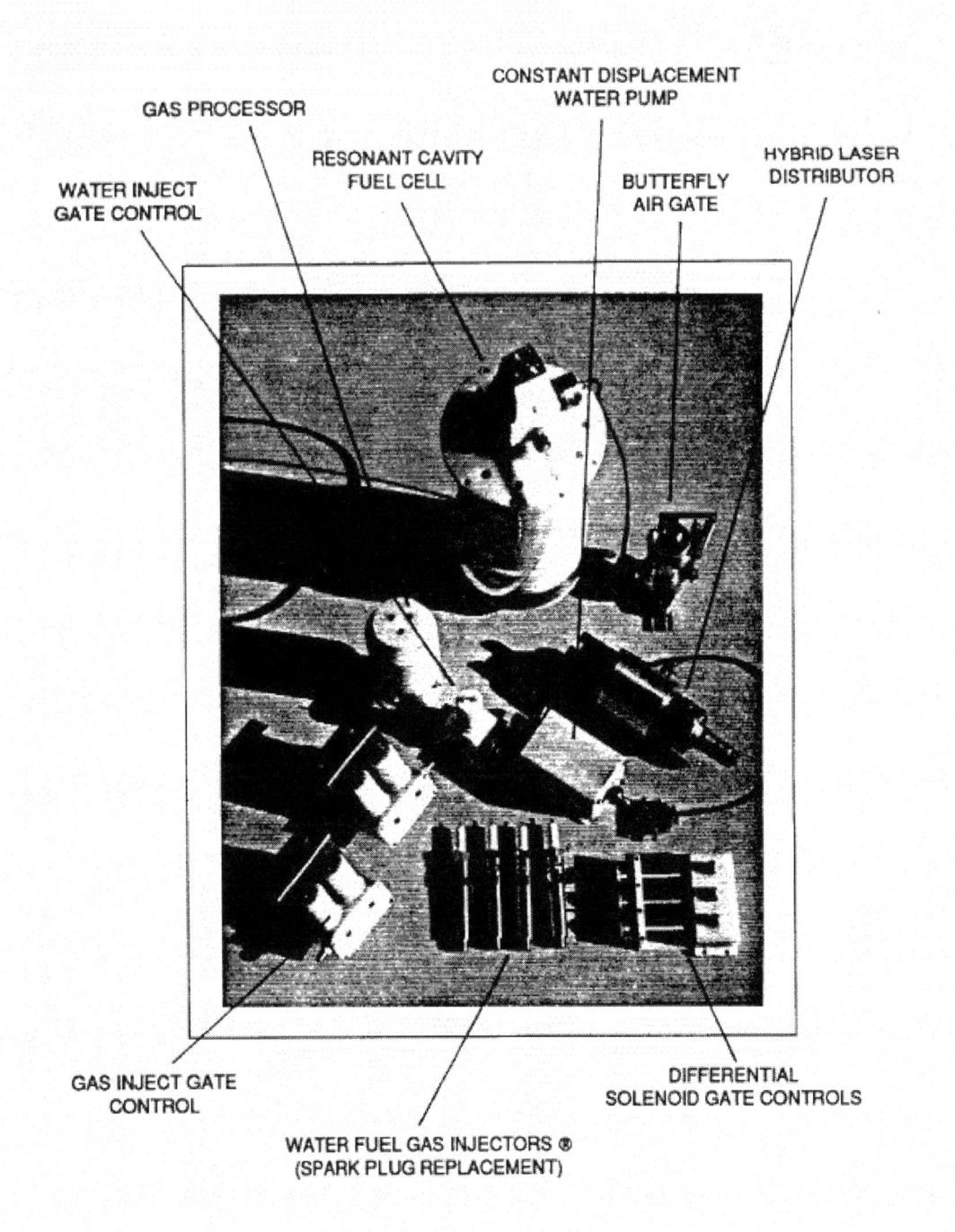

WATER FUEL GAS INJECTION SYSTEM

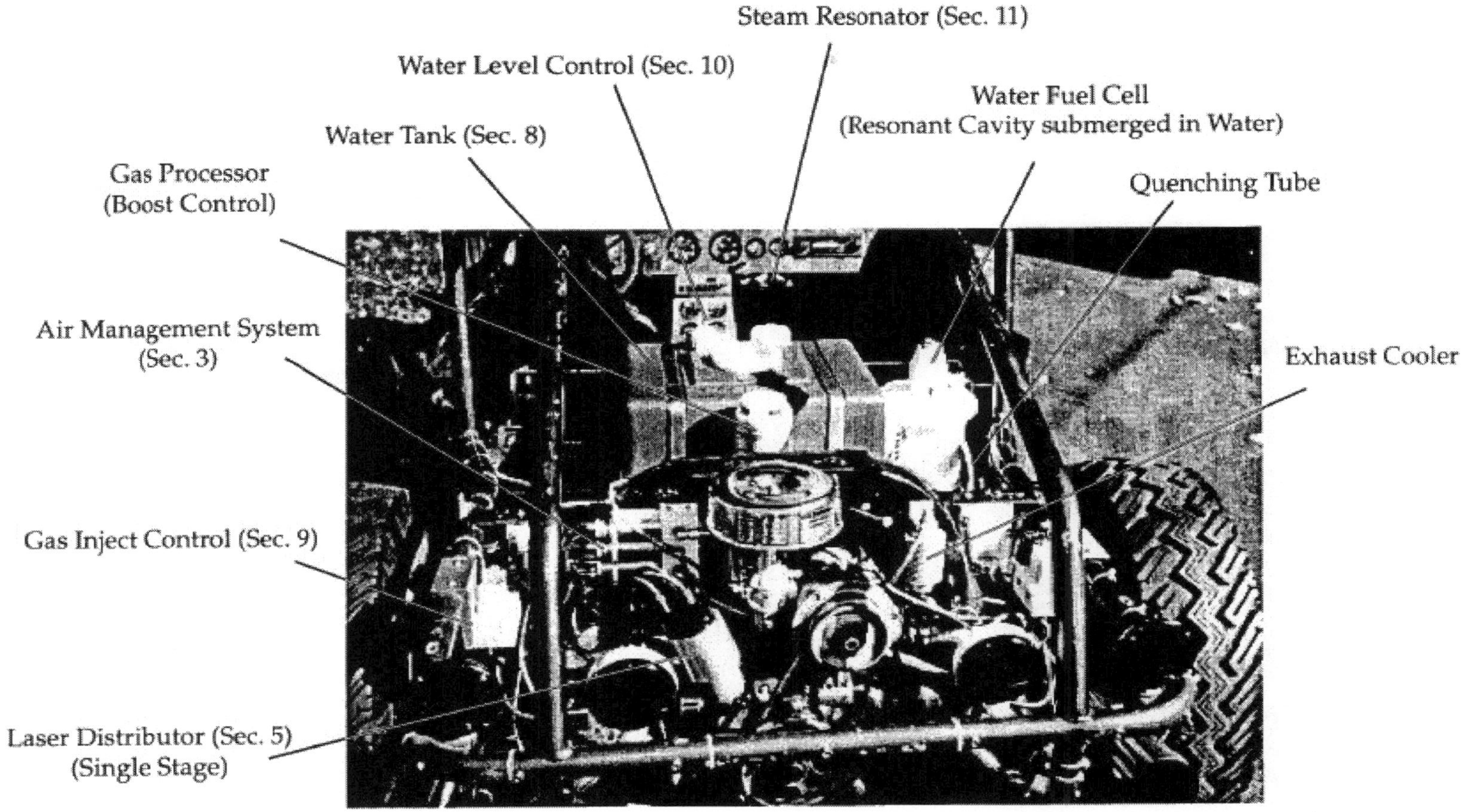

Gas Processor
(Boost Control)

Air Management System
(Sec. 3)

Gas Inject Control (Sec. 9)

Laser Distributor (Sec. 5)
(Single Stage)

Water Tank (Sec. 8)

Water Level Control (Sec. 10)

Steam Resonator (Sec. 11)

Water Fuel Cell
(Resonant Cavity submerged in Water)

Quenching Tube

Exhaust Cooler

Water Fuel-Gas Injection System
(Component Placement prior to debugging stage)
Dune Buggy: Rear View

WFC Project 422 DA

Memo WFC 423 DA

Water Fuel Cell

Water Fuel Injection System ®

Water Fuel Injector System ® processes and converts water into a useful hydrogen fuel on demand at the point of gas ignition.

The Water Injector System ® is design variable to be retrofitted by replacing fossil-fuel injectors-ports affixed to conventional jet engines, heating system, rocket engines, even replacing internal combustion engine spark plugs.

The Water Fuel Management (WFMS) System is a digital computer logic control system which systematically activates the Water Fuel Injection System ® in the following way...using water as fuel.

Section 4

GAS PROCESSOR

STANDARD 1600CC
I.C. ENGINE

WATER FUEL INJECTORS
(SPARK PLUG REPLACEMENT)

WATER TANK

WATER FUEL MANAGEMENT (WFMS)
SYSTEM MODULE

LASER ACCELERATOR
CONTROL MODULE

WATER FUEL INJECTION SYSTEM
(Component Placement prior to debugging stage)

Dune Buggy: Rear Side View

WFC PROJECT 423 DA

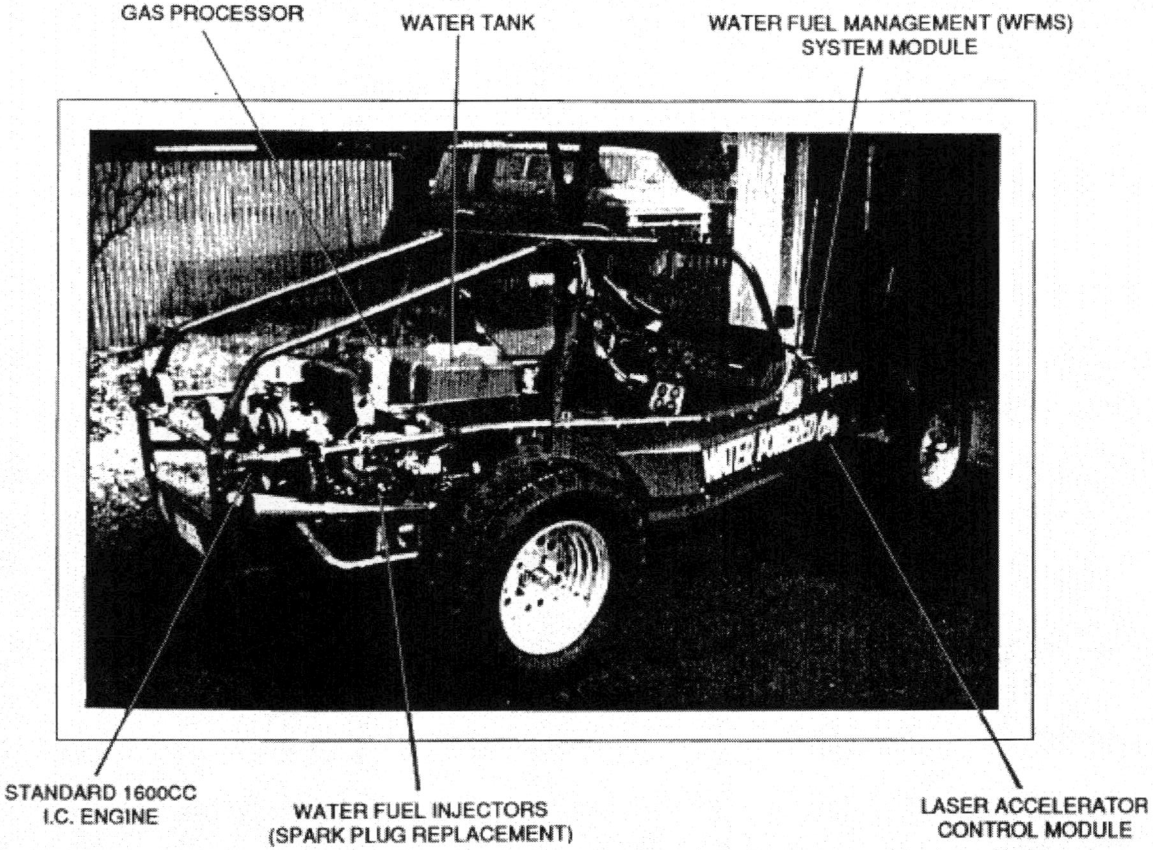

STANDARD 1600CC
I.C. ENGINE

GAS PROCESSOR

WATER FUEL INJECTORS
(SPARK PLUG REPLACEMENT)

WATER TANK

WATER FUEL MANAGEMENT (WFMS)
SYSTEM MODULE

WATER FUEL INJECTION SYSTEM
(Component Placement prior to debugging stage)

LASER ACCELERATOR
CONTROL MODULE

Dune Buggy: Rear Side View

WFC PROJECT 423 DA

WATER INJECT GATE CONTROL

GAS PROCESSOR

**WATER FUEL INJECTORS ®
(SPARK PLUG REPLACEMENT)**

BUTTERFLY AIR GATE

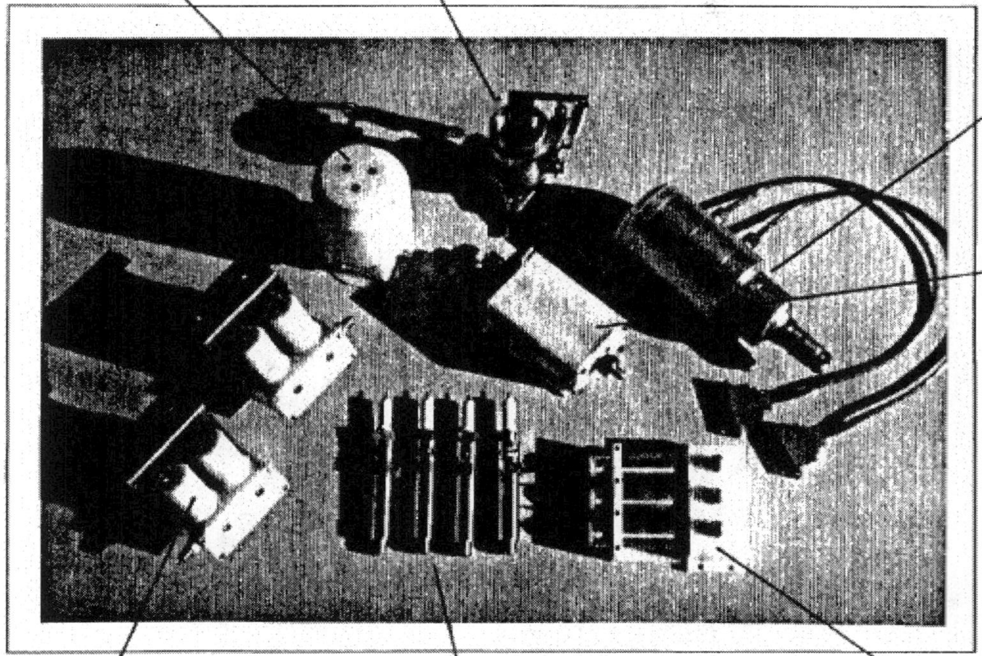

WATER FUEL INJECTION SYSTEM

Retrofit Kit

DIFFERENTIAL SOLENOID GATE CONTROLS

CONSTANT DISPLACEMENT WATER PUMP

**DUAL LASER DISTRIBUTOR
(INTERCHANGEABLE WITH
HYBRID LASER DISTRIBUTOR)**

WFC PROJECT 423 DA

VOLTAGE INTENSIFIER CIRCUIT COIL ASSEMBLY

WATER FUEL INJECTORS (SPARK PLUG REPLACEMENT)

DIFFERENTIAL SOLENOID GATE CONTROLS

WATER INJECT CONTROL

HYBRID LASER DISTRIBUTOR (INTERCHANGEABLE WITH DUAL STAGE LASER DISTRIBUTOR)

WATER FUEL INJECTION SYSTEM
(Component Placement prior to debugging stage)

Dune Buggy: I.C. Engine Side View

WFC PROJECT 423 DA

WFC PROJECT 423 DA

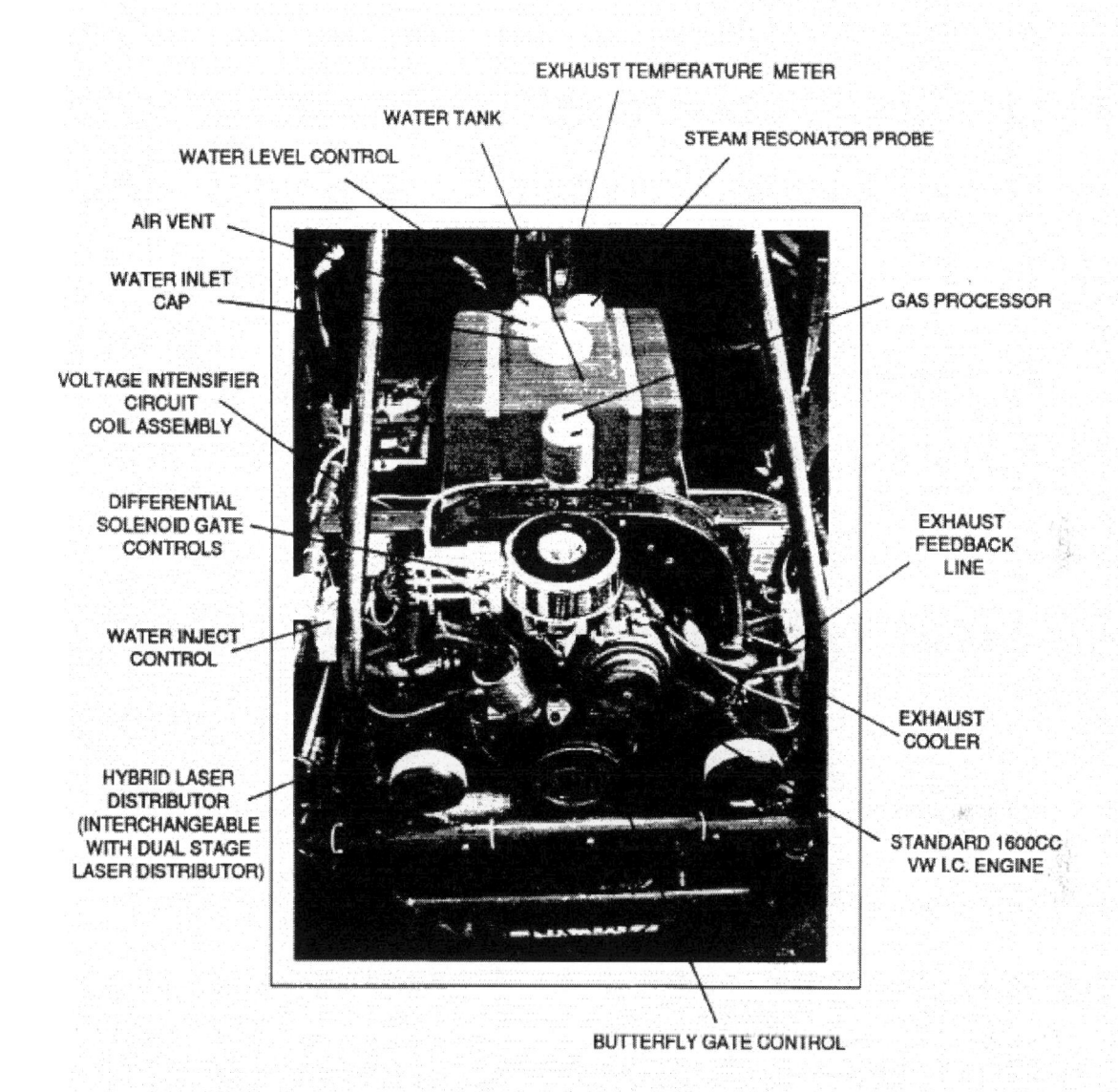

- EXHAUST TEMPERATURE METER
- WATER TANK
- WATER LEVEL CONTROL
- STEAM RESONATOR PROBE
- AIR VENT
- WATER INLET CAP
- GAS PROCESSOR
- VOLTAGE INTENSIFIER CIRCUIT COIL ASSEMBLY
- DIFFERENTIAL SOLENOID GATE CONTROLS
- EXHAUST FEEDBACK LINE
- WATER INJECT CONTROL
- EXHAUST COOLER
- HYBRID LASER DISTRIBUTOR (INTERCHANGEABLE WITH DUAL STAGE LASER DISTRIBUTOR)
- STANDARD 1600CC VW I.C. ENGINE
- BUTTERFLY GATE CONTROL

WATER FUEL INJECTION SYSTEM
(Component Placement prior to debugging stage)

Dune Buggy: Rear View

RE: Water Fuel Injection SystemMemo WFC 423 DA

Water Fuel Injection System ®

      WFC Hydrogen Gas Management System is ideally suited as a retrofit energy system to both reciprocating (rotary piston engine) and turbine jet engines associated with the aviation industry...but in different ways: Reciprocating WFC fuel-kits can be similar to car design (340) of Figure (3-38) of WFC (422 DA) ; Whereas, Water Fuel Injector Kit (10) of Figure (4-1) can be alternately be used as a self-contained Fuel-unit having no pressurized vessel which converts water directly into thermal explosive energy (gtnt) on demand , as illustrated (10) of Figure (4-1) as to Figure (40) of Figure (4-2).

      Operationally, Water Fuel injector assembly (10) of Figure (4-1) as to (40) of Figure (4-2) performs several function simultaneously to produce thermal explosive energy-yield (gtnt) (16) on demand:

      First water mist (47) of Figure (4-4) is injected into fuel-mixing chamber (35) of Figure (4-5) by way of water spray ports (41a xxx 41n) of Figure (4-4); Secondly, ionized air gases (46a xxx 46n) of Figure (4-4) (laser primed ambient air gases having missing electrons) produced by Ambient Air Ionizer (80) of Figure (4-6) as to Figure (4-1) and non-combustible gases (45) of Figure (4-4) are intermixed with expelling water mist (47a xxx47n) to form Water-fuel mixture (48) by way of gas mixing disc (34) of Figure (4-5) as to (30) of Figure (4-2); thirdly, the resultant moving Water-Fuel mixture (48) of Figure (4-5) enters into Voltage Igniter Stage (180) of Figure (4-5) and exposed to high intensity voltage fields (33 / 36) (typically 2,000 volts or above @ 10 Khz or above) of opposite electrical polarity (E7 / E8) ...which, in turn, not only performs electrical polarization process (160) of Figure (3-26) undergoing Dielectric Resonant (240) of Figure (3-31); but, also, sets up and triggers Hydrogen Fracturing Process (390) of Figure (3-42) as to Figure (3-6) under control state (on demand) via electrical-static spark ignition (49 / 51) of Figure (4-5)....releasing thermal explosive energy (gtnt) (16) passing beyond gas exit port (32) of Figure (4-5), as further illustrated in Figure (4-2) as to Figure (4-1).

      To ensure proper energy-flame projection and subsequent energy-flame stability, constant displacement water pump (170) causes and allows ionized ambient air gases (46), non-combustible gases (45), and water (47) to be displaced under static pressure up to and beyond 125 lbs psi, respectively.

RE: Water Fuel Injection System   Memo WFC 423 DA

Energy-Flame density is enhanced and sustained by causing ionized gases (46a xxx 46n) of spray port (42) to be deflected into liquid spray path (41), together water mist (47) and ionized air gas (46) are, now, directed toward and deflected through non-combustible gas spray path (43)...producing uniformed water-fuel mixture (48), as illustrated in Figure (4-5).

Energy-Flame temperature is regulated by controlling the volume flow-rate of each fluid-mediums (47 / 45/ 46) in direct relationship to applied voltage intensity (33 /36), as further illustrated in Figure (4-2) as to Figure (4-5). To elevate Energy-flame-temperature still further, simply increase fluid-displacement (46/47) while maintaining or reducing the volume flow rate of non-combustible gases (45) during an increase of applied voltage amplitude (V0 xxx Vn) of Figure (4-2) as to Voltage Intensifier Circuit (110) of Figure (4-9) and Electron Extraction Circuit (120) of Figure (4-10). To lower Energy-flame temperature simply increase the amount of non-combustible gases (45a xxx) or reduced the fluid flow rate (45 / 46/ 47)) uniformly while lowering pulse voltage amplitude (xxx V0). To establish a predetermined or given Energy-flame temperature adjust fluid-medium (45 / 46 / 47)) and applied voltage amplitude (V0 xxx) independent of each other to obtain the desired results.

The resultant energy-flame pattern is further maintained by allowing the ignited, compressed, and moving gases (29) of Figure (4-5) to be projected to, pass through and beyond nozzle-port (32) under pressure due to gas expansion caused by thermal gas ignition.

Voltage Igniter Stage (180) of Figure (4-5) as to Voltage Intensifier Circuit (110) Figure (4-9) as to Extraction Circuit (120) of Figure (4-10) performs several functions simultaneously to initiate and trigger thermal explosive energy-yield (gtnt) (16) beyond normal gas burning levels:

Water droplets (28a xxx 28n) escaping from spray-mist (47) and exposed to high intensity voltage fields of opposite polarity 33/ 36) are stimulated to undergo Electrical Polarization Process (160) of Figure (3-26)...which not only separates and splits the unlike atoms of the water molecule but also causes the unlike atoms (hydrogen atoms 77a / 77b and oxygen atom 76) to experience electron ejection (230) of Figure (3-30) as to (71) of Figure (4-10) since voltage intensifier circuit (110) of Figure (4-9) inhibits and prevents electron flow to enter into gas ignition process (180), as further illustrated in Figure (4-8).

Stanley A. Meyer

RE: Water Fuel Injection System												Memo WFC 423 DA

    The newly liberated water molecule atoms (oxygen 76 and hydrogen atoms 77a / 77b) immediately interact with laser primed ionized ambient air gases (7a xxx 7n of Figure 1-15) (see WFC memo 420) to cause the resultant highly energized and mass destabilized combustible gas atoms (93a xxx 93n) of Figure (4-10) to perform Hydrogen Fracturing Process (80) of Figure (4-9) when electrostatic force (14/16) thermally ignites (kinetic agitation) destabilized water-fuel mixture (93a xxx 93n) under gas compression...preventing the formation of the water molecule during thermal gas ignition....satisfying Energy Gas Detonation Equation .

(Eq 18)

$$E_{in} \stackrel{gas}{\underset{det}{=}} M_d C^2 \quad \text{Thermal Explosive Energy (gtnt)}$$

Which states

    That, whenever the mass-size of a combustible gas atom is decreased (M_d), thermal explosive energy-yield (gtnt) is increased (E_{in}) during thermal gas combustion (Gas // Detonation.), as so illustrated in (100) Figure (4-8) as to (90) of Figure (4-7).

    Incoming ambient air gases (5a xxx 5n) become laser primed and ionized when passing through Ambient Air Ionizer (Gas Processor) (80) of Figure (4-6) as to (10) of Figure (4-1) since electron extraction circuit (120) of Figure (4-10) not only captures and consumes ejected electrons (7a xx 7n) of Figure (4-8); but, also prevents electron flow into destabilizing gas process (180), as so illustrated in Figure (4-5).

    In terms of performance reliability and safety, ionized air gases (46a xxx 46n) and liquid water (47a xxx 47n) do not become energy activated (volatile) until water-fuel mixture (48) reaches Voltage Igniter Stage (180). Injected non-combustible gases (45a xxx 45n) retards and controls the combustion rate of the Hydrogen Fracturing Process (100) of Figure (4-8) during gas-ignition.

    In other or alternate applications, laser primed ionized liquid oxygen (68) of Figure (1-21) (see WFC memo 420) and laser primed liquid hydrogen (69) of Figure (1-21) stored in separate fuel-tanks can be used in place of fuel-mixture (48); or, liquefied ambient air gases (6) alone with water-source (8) can, also, be substituted as a fuel-source (48) to trigger Hydrogen Fracturing Process (100). Additional WFC Injector Assemblies (20) of Figure (4-1) are arranged in cluster

RE: Water Fuel Injection System   Memo WFC 423 DA

array (20a xxx 20n) to increase energy-yield output (16a xxx 16n) of Figure (4-12/4-13/4-14).

WFC injector assembly (10) of Figure (4-1) as to (30) of Figure (4-2) is design variable to be retrofitable by replacing fossil-fuel injector ports affixed to jet engines (see Figure 4-13), heating systems (Figure 4-12), rockets engines (Figure 4-14), or even car spark plugs (130) of Figure (4-11) which simply uses Water Fuel management (WFMS) system fluid- metering system (40) to control gas ignition (16), as illustrated in (40) of Figure (4-2). Sequential pulsing of Water Fuel Injector (20/30) of Figure (4-1) as to (40) of Figure (4-2) is system activated by Pulse Gate Valve (190) of Figure (4-1) to further control a predetermined energy-flame (16).

In essence, then, the Water Fuel Injector system (40) simply processes and converts water into a useful hydrogen fuel on demand at the point of gas ignition...thereby, co-equally or superseding fossil-fuel safety standards...especially when ionized ambient air gases (46a xxx 46n) and non-combustible gases (45a xxx 45n) are intermixed with water supply (47) prior to entering Water Fuel Injector Plug (20 / 30), as illustrated in (40) of Figure (4-2) as to (10) of Figure (4-1).

RE: Water Fuel Injection System   Memo WFC 423 DA

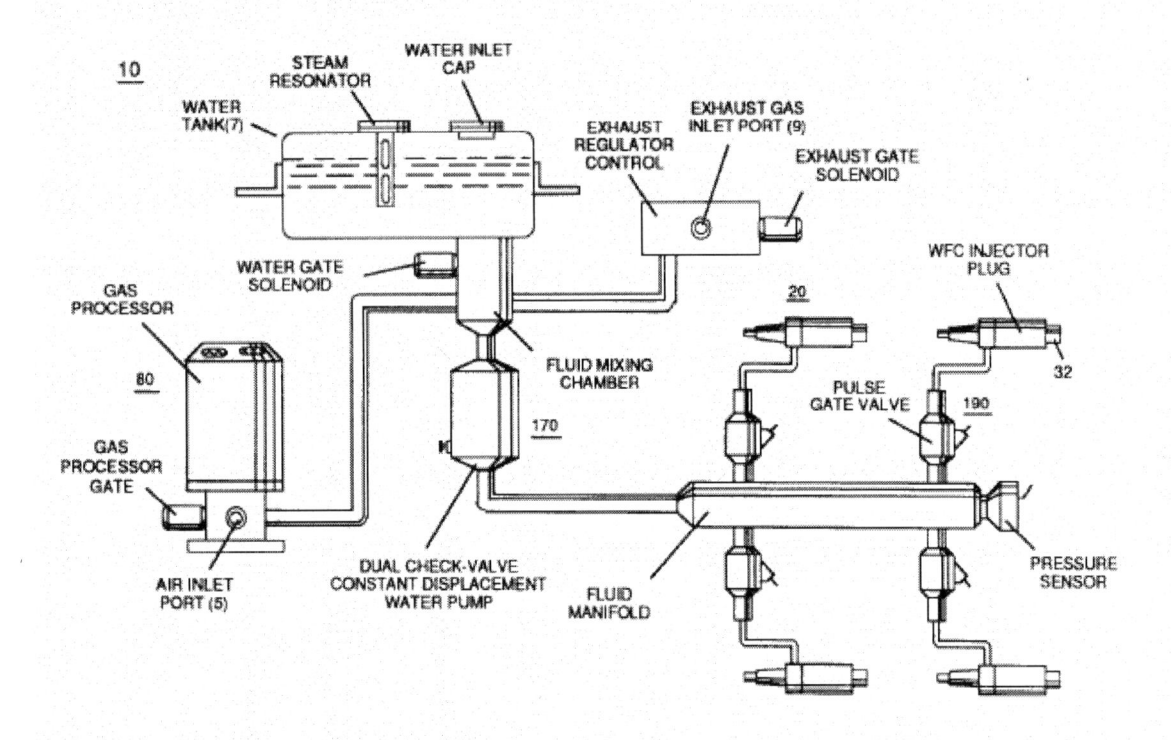

FIGURE 4-1: WATER FUEL INJECTOR SYSTEM

RE: Water Fuel Injection System   Memo WFC 423 DA

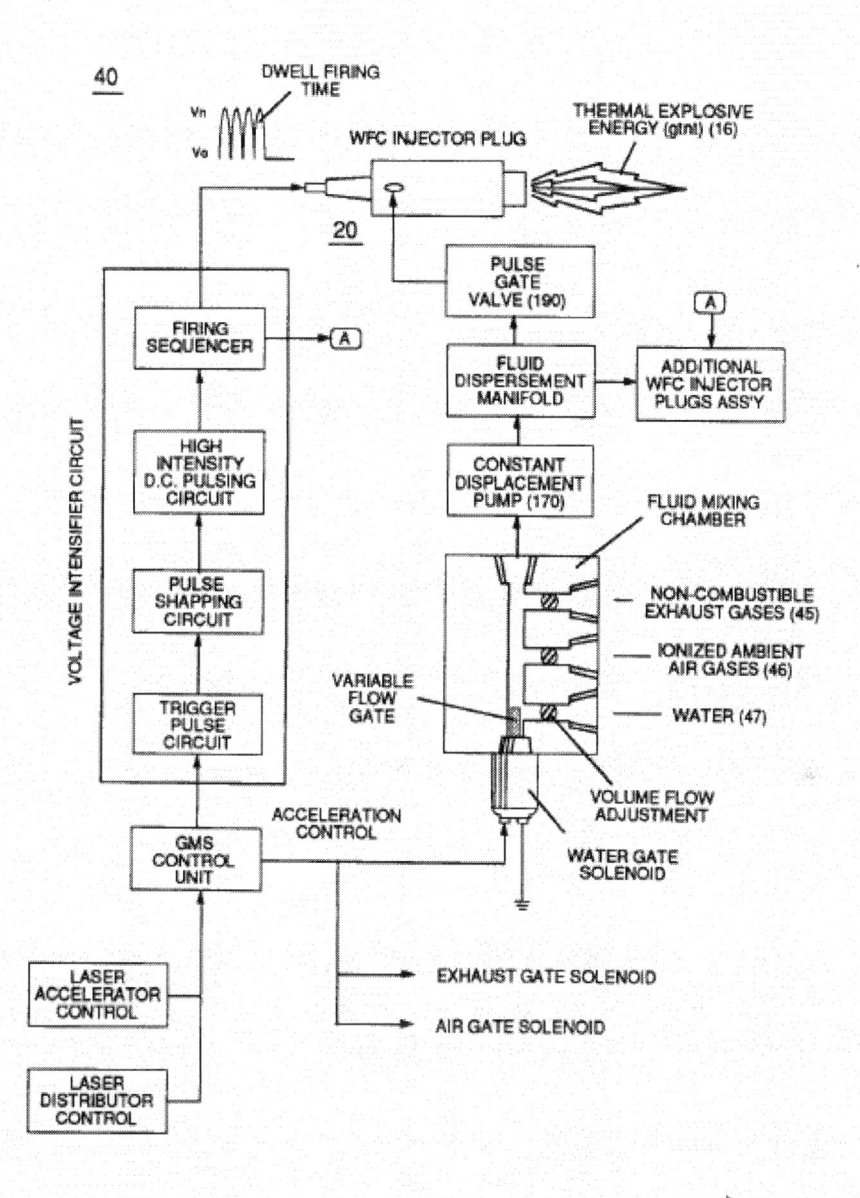

FIGURE 4-2: WATER FUEL MANAGEMENT (WFIS) SYSTEM

Stanley A. Meyer   4-6

RE: Water Fuel Injection System   Memo WFC 423 DA

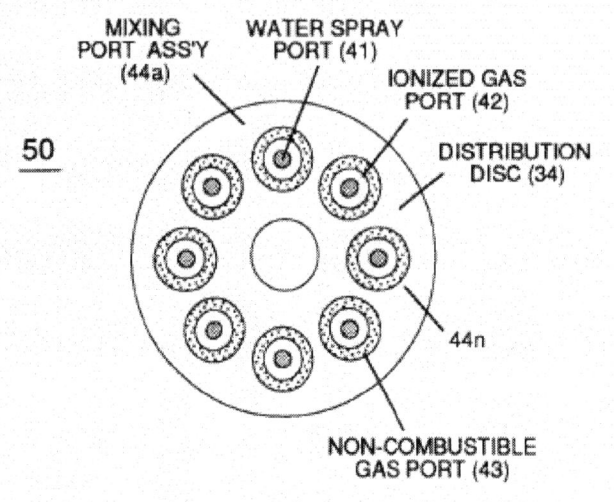

FIGURE 4-3: SPRAY DISTRIBUTION DISC

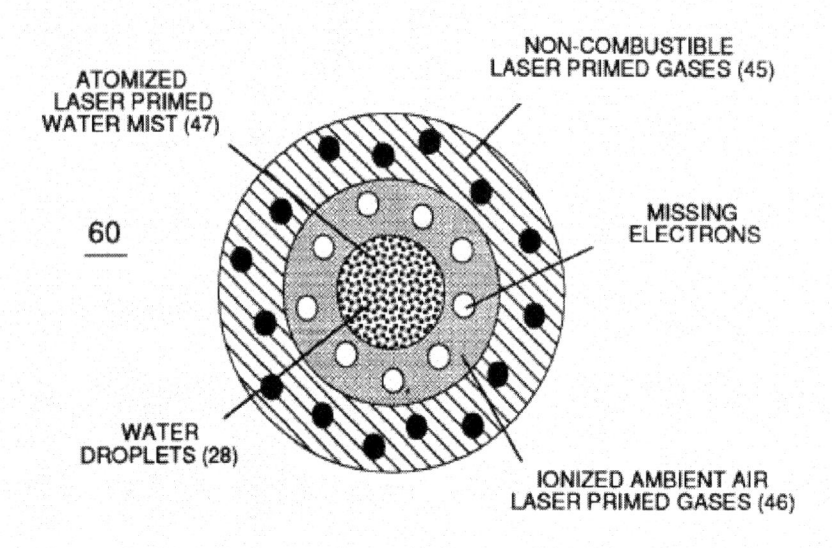

FIGURE 4-4: SPRAY PATTERN

RE: Water Fuel Injection System   Memo WFC 423 DA

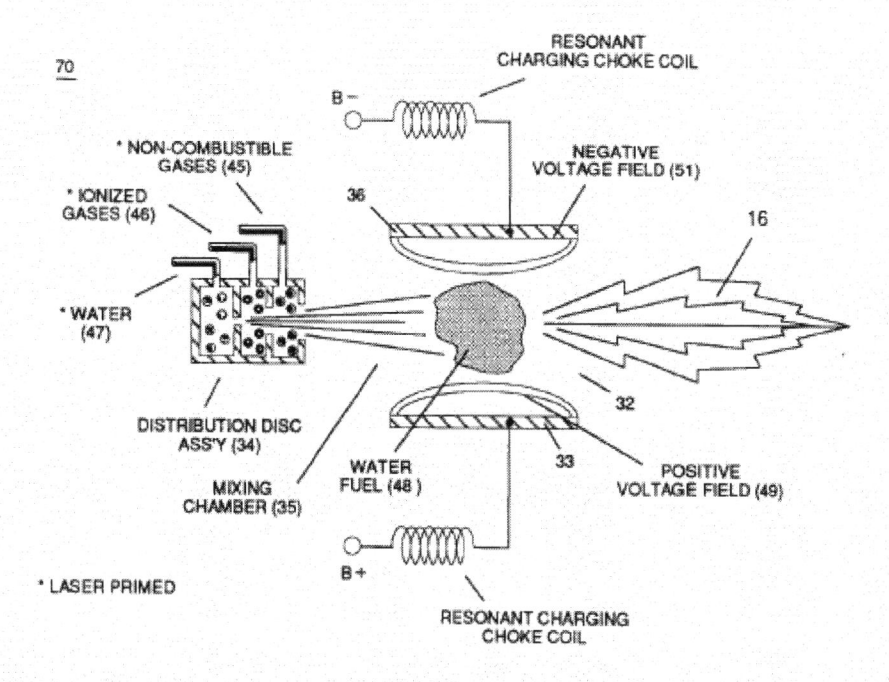

FIGURE 4-5: VOLTAGE TRIGGERING

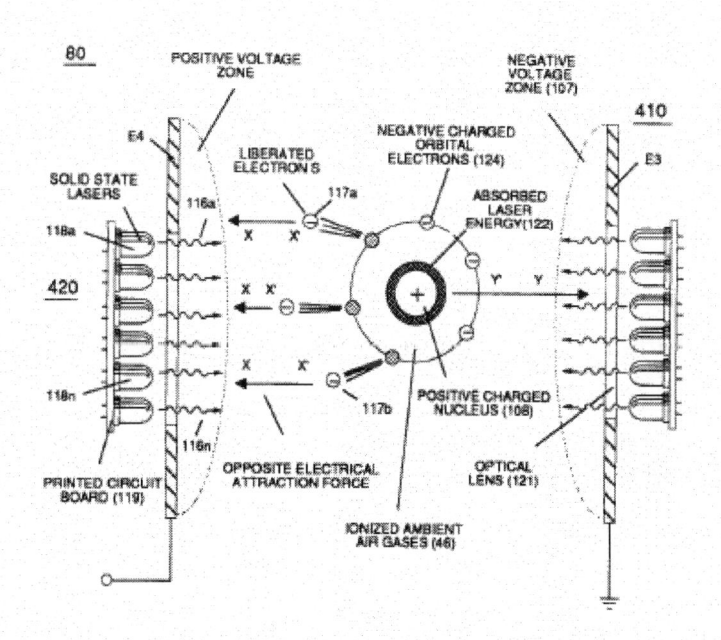

FIGURE 4-6: AMBIENT AIR IONIZER

Stanley A. Meyer

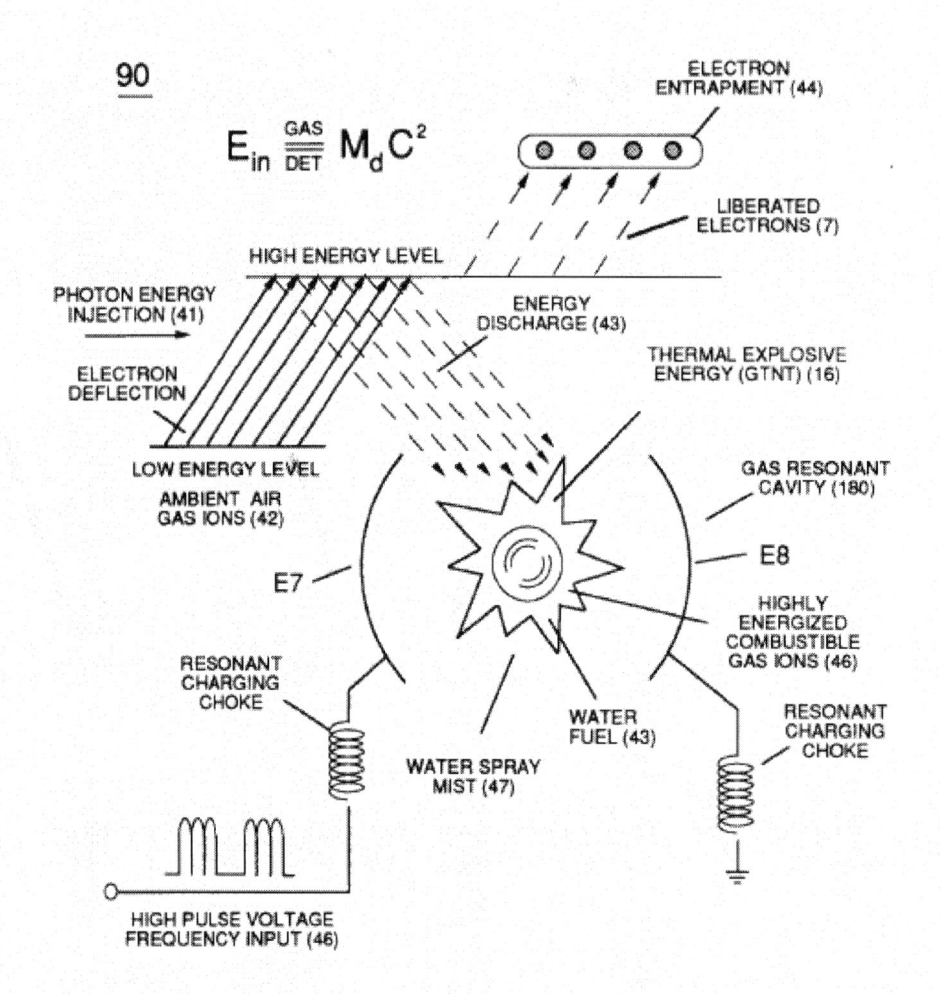

FIGURE 4-7: VOLTAGE IGNITER STAGE

RE: Water Fuel Injection System					Memo WFC 423 DA

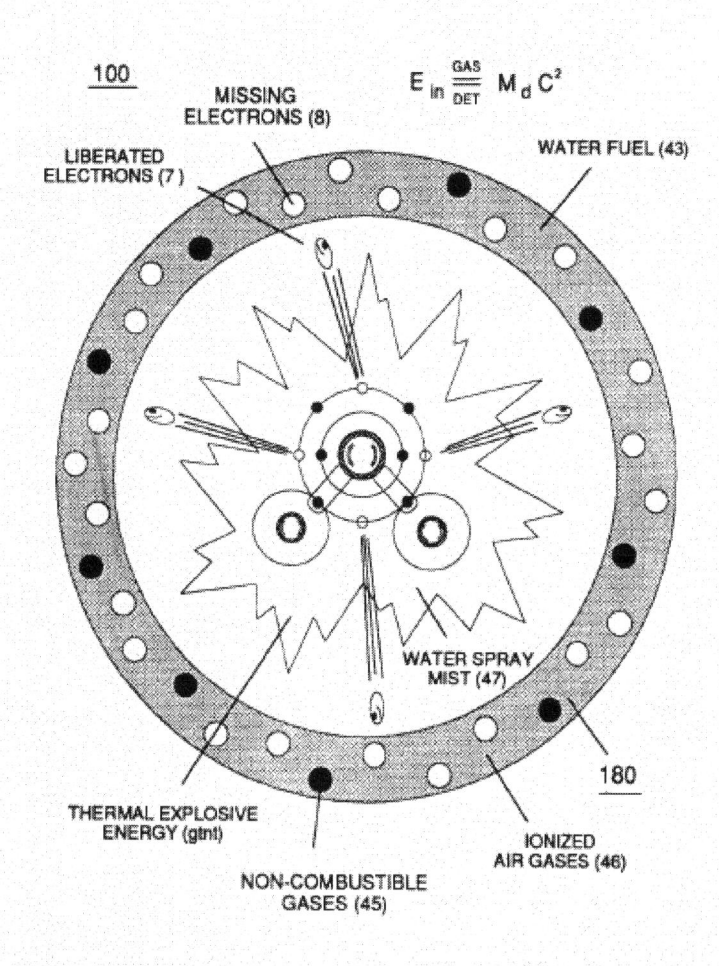

$$E_{in} \overset{GAS}{\underset{DET}{\equiv}} M_d C^2$$

- 100
- MISSING ELECTRONS (8)
- LIBERATED ELECTRONS (7)
- WATER FUEL (43)
- WATER SPRAY MIST (47)
- 180
- THERMAL EXPLOSIVE ENERGY (gtnt)
- NON-COMBUSTIBLE GASES (45)
- IONIZED AIR GASES (46)

FIGURE 4-8: GAS IGNITION STAGE

Stanley A. Meyer					4-10

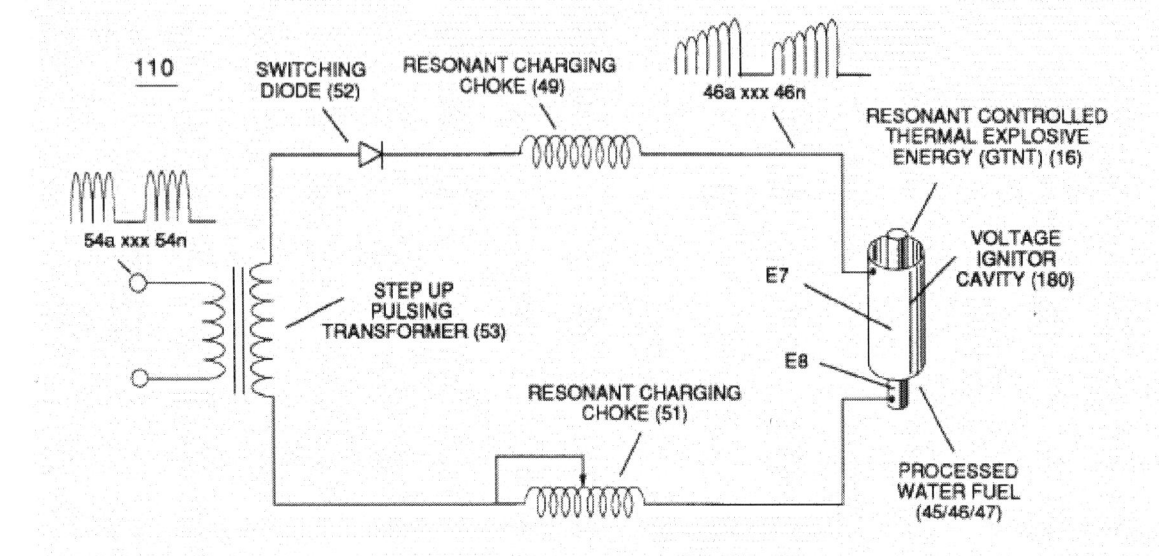

FIGURE 4-9: VOLTAGE INTENSIFIER CIRCUIT

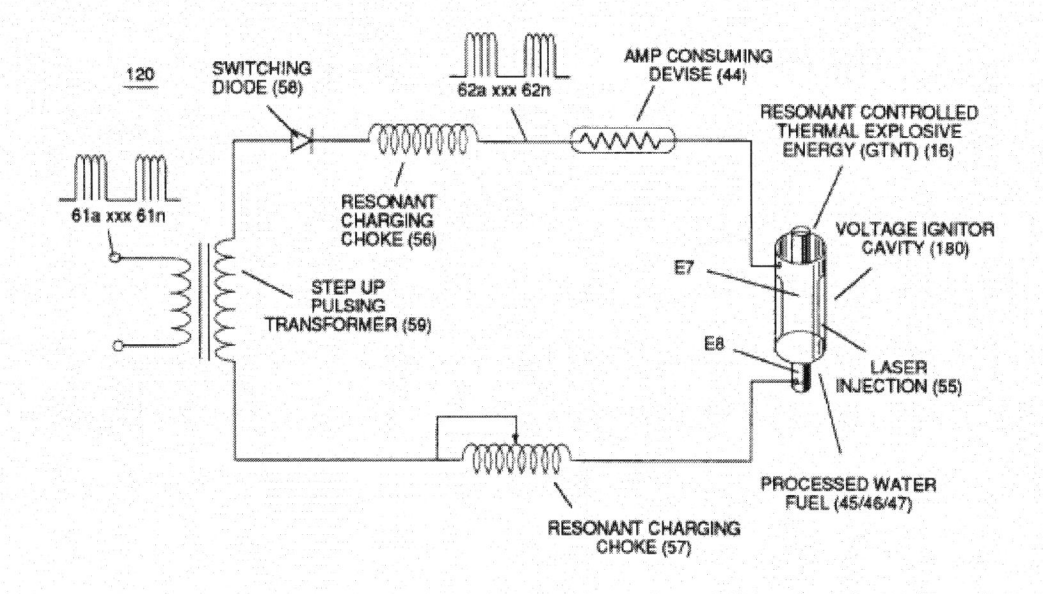

FIGURE 4-10: ELECTRON EXTRACTION CIRCUIT

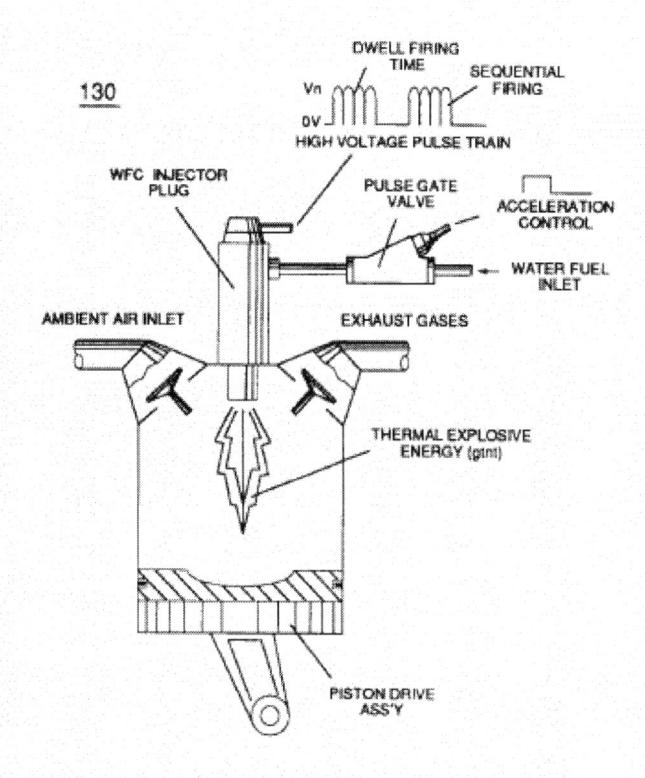

FIGURE 4-11: INTERNAL COMBUSTION ENGINE RETROFIT

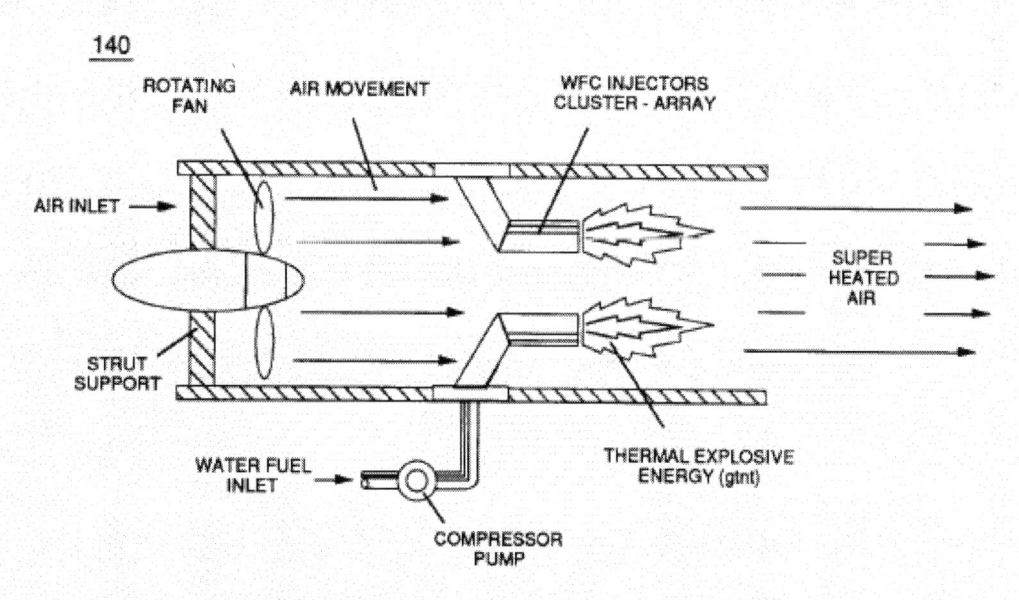

FIGURE 4-12: FURNACE RETROFIT

RE: Water Fuel Injection System Memo WFC 423 DA

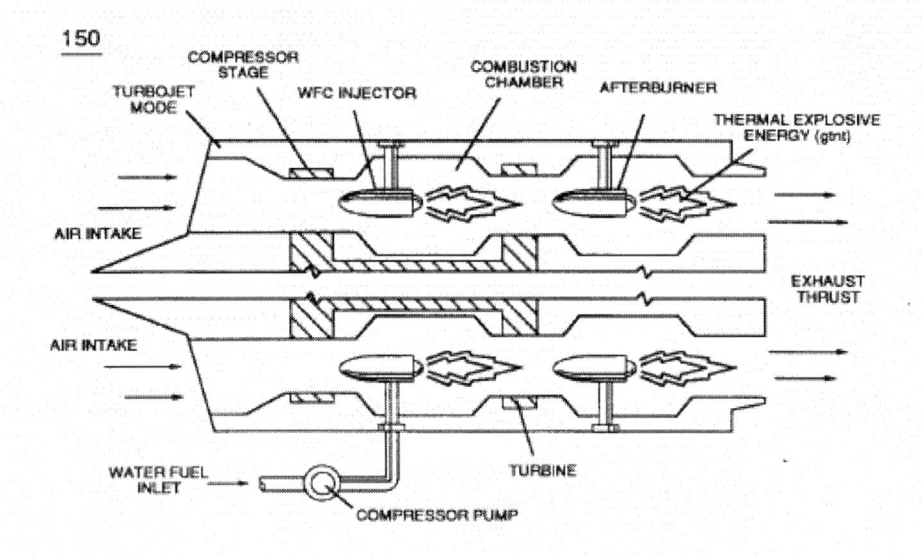

FIGURE 4-13: JET ENGINE RETROFIT

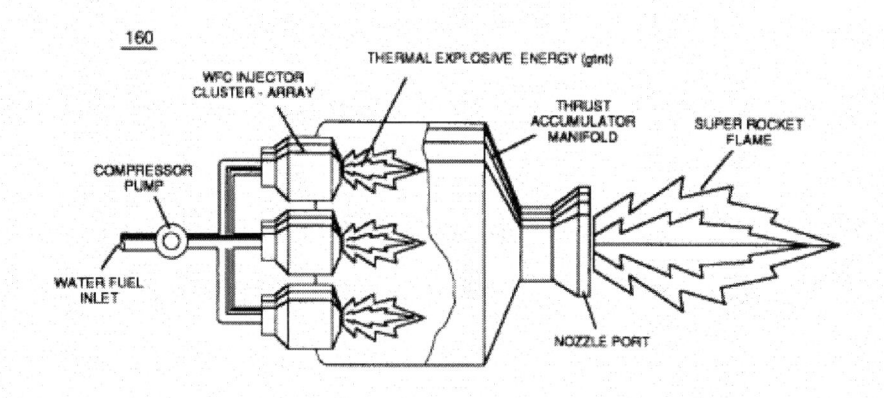

FIGURE 4-14: ROCKET ENGINE RETROFIT

Memo WFC 424

Atomic Energy Balance of Water

Where does the energy come from if voltage is being used to separate the the water molecule into it's component gases for hydrogen gas utilization ?

Is it possible to attenuate the electrical fields of the combustible gas atoms to increase there energy level prior to gas ignition ?

By what manner can the Atomic Energy Balance of the combustible gas atoms be changed to increase thermal explosive energy-yield ?

Is Electrical Pressure of opposite polarity only required to oscillate the Energy Aperture of combustible gas atoms to allow the use of Universal Energy ?

Voltage stimulation of the water molecule by way of the Electrical Polarization Process, now , sets up and triggers the Hydrogen Fracturing Process for energy enhancement ...and is performed in the following way:

Section 5

RE: Atomic Energy Balance of Water   Memo WFC 424

Atomic Energy Balance of Water

Using Water as Fuel

Abstract

The Atomic Energy Balance of Water is activated and performed in a sequence of events in an instant of time. The Hydrogen Fracturing Process simply triggers and releases atomic energy from natural water by retarding and slowing down the reformation of the water molecule being subjected to sub-critical state during thermal gas ignition. The Voltage Intensifier Circuit brings on the Electrical Polarization Process that switches off the covalent bond of the water molecule without amp influxing. Energy Pumping Action undergoing "Resonant Propagation" of opposite electrical stress oscillates the "Energy Aperture" of the combustible gas atoms to increase the atomic energy level of the combustible gas atoms prior to gas ignition. The Electron Extraction Process ionizes the highly energized combustible gases to decrease atomic mass while applied Voltage Pulse-Frequency of opposite electrical polarity initiates the triggering process once maximum voltage deflection is achieved ... releasing Thermal Explosive Energy (gtnt) beyond normal gas burning levels. The energy contained in a gallon of water exceeds 2.5 million barrels of oil when equated in terms of atomic energy. Water, of course, is free, abundant, and energy recyclable.

Energy Pumping Action

Once unlike atoms of the Water Molecule (210) separates into it's component gases by way of Electrical Polarization Process (160), the newly liberated and subsequently free floating hydrogen (77 a/b), oxygen (76), and dissolved air gases (97) (see WFC memo 420 DA Figure 3-25) submerged in water bath (68) are further subjected to pulse-voltage stimulation (SS' / RR ' ~ TT ' / UU ')...exerting Electrical Pressure of opposite polarity on the atomic level, as illustrated in Figure (5-1) as to Figure (5-2) and Figure (5-8).

Applied voltage fields (66-E1 / 67-E2) causes and forms both opposite attraction forces (ST - ST') and (RU - RU ') across exposed Hydrogen Atom (77) since the negative charged hydrogen electron (1) is attracted and deflected toward stationary positive voltage field (66); while, simultaneously, positive charged hydrogen proton (3) (Hydrogen Nucleus) is attracted and moved in opposite direction toward stationary voltage field (67)...causing elongation of the orbital path of moving electron (1)...changing the time share rate of the electron...applying and superimposing

Stanley A. Meyer

electrical tension (4) onto opposite attraction force (AA') that exist between the negative charged orbital electron (1) and positive charged Proton (3)...which, in sequence, applies and superimposes electrical tension (5) unto opposite attraction forces (ZZ') that exists and occupies space between proton particles masses (6a xxx 6n) and Energy Aperture (7).

Interlocked together, the resultant "Electrical Tension of Forces" (4 / 5) is directly related to Voltage Intensity (ST - ST' / RU - RU') of applied voltage fields (66/67) which is variable as to applied voltage amplitude (Vo xxx). Increasing voltage amplitude (xxx Vn) still further increases electrical tension (4 / 5) being applied to Energy Aperture (7) or vice versa. This established increase or decrease in Electrical Pressure (4 / 5), now causes Energy Aperture (7) to either enlarge or become smaller as to applied voltage amplitude (Vo xxx Vn), respectively. Once applied voltage pulse of opposite polarity (66/67) is terminated during pulse off-time, then Energy Aperture (7) automatically adjusts to maintain a given energy level since each aperture oscillation (aperture expansion) emits a discrete amount of Universal Energy (9) into proton (3) via energy pathway (ZZ'), as illustrated in Figure (5-1).

Repetitive pulsing of applied voltage fields (66/67), now, oscillates Energy Aperture (7) to emit even a greater amount of Universal Energy (9a xxx 9n) into the energy spectrum of the proton nucleus (3) to be absorbed...thereby, increasing the energy level of hydrogen atom still further....deflecting hydrogen electron (1) to a higher energy state, as illustrated in (520) of Figure (5-3). The resultant energy state of increase is directly related to applied pulse voltage frequency at a predetermined Voltage level.

Likewise, liberated oxygen atom (76) and other dissolved air gases (97) submerged in water bath (68) undergo and experience similar Energy Pumping Action (520) (aperture oscillation) of there respective Energy Aperture (11a xxx 11n) when exposed to voltage stimulation (66/67), as further illustrated in Figure (5-2). In this phase of application, opposite attraction force (BB') provides a energy transfer path (12) to each respective proton (3a xxx 3n) from energy aperture (11) which is centrally formed during proton grouping...establishing nucleus (14) of Figure (5-2).

Resonant Propagation

These highly energized and liberated water bath atoms (76/77/97), now, causes Resonant Action to occur at a progressive rate during continued voltage stimulation...giving way to the

RE: Atomic Energy Balance of Water　　　　　　　　　　　　　　　　Memo WFC 424

following operational parameters of hydrogen gas production for energy utilization from natural water:

Resonant Action (21) (point of particle oscillation) occurs when applied pulse voltage frequency (46) of Figure (5-5) is adjusted to "tune-in" to the Dielectric Resonance of water via Voltage Intensifier Circuit (60) of Figure (5-3); whereas, applied voltage amplitude (Vo xxx Vn) which is independent of Resonance Frequency is adjusted to cause water bath atoms to momentarily enter into Liquid-to-gas ionization state....ejecting negative charged electrons....forming positive charged atoms having missing electrons ...forming negative charged atoms by electrons capture, as illustrated in (520) in Figure (5-3) as to Figure (5-4A) and Figure (5-4B).

Compounding Action (22) (deflection of electrical charged particles) by way of voltage stimulation aids Resonant Action by superimposing particle impact onto to the Electrical Polarization Process (160).

Resonant Action (21), Compounding Action (22), Laser Injection (17), and Energy Pumping Action (520), now, allows the production of hydrogen and oxygen gases from water in geometrical progression to set up Hydrogen Fracturing Process (90), as illustrated in Figure (5-4C) as to Figure (5-5).

Triggering Process

As water mist (68a xxx 68n) is injected and metered into Resonant Cavity (180) of Figure (5-5) (Water Fuel Injector) under pressure, applied Resonant Pulse Voltage (46) performs several functions sequentially in a instant of time: converts water mist (68) into it's component gases hydrogen (77 a/b), oxygen (76), and ambient air gases (97)); momentarily ionizes the liberated gases by way of electron ejection (230), and thermally ignites ionized combustible gas-mixture under "Electrostatic Pressure" that directly attenuates Energy Apertures (7a xxx 7n) (520)....releasing thermal explosive energy (gtnt) beyond normal gas burning levels on demand, as illustrated in Figure (5-5). Once a quantum amount of thermal explosive energy (gtnta xxx gtntn) is released then the combustible gases return to stable state by forming de-energized water mist (531), as shown in (530) of Figure (5-6).

Stanley A. Meyer

RE: Atomic Energy Balance of Water   Memo WFC 424

Energy Recycling Spectrum

Once expelled into Earth Atmosphere (541), the energy spent (partly de-energized water molecule atoms) water mist (531) is re-exposed to Sun Energy (534) to allow and cause Photon Energy Absorption Process (532) of Figure (5-7) that, now, re-vitalizes and returns de-energized Water Molecule Atoms (531a xxx 531n) to Stable Energy State (538) when spent water mist (531) undergo Water Evaporations Process (531) to form Moisture Cloud Formations (533) which, when and after formed, are continually exposed to incoming Sun Rays (539a xxx 539n) until Rain Droplets (535) return newly re-energized water molecules (538a xxx 538n) to Earth Water Supply (536)...completing Open Ended Energy Recycling Spectrum (530) for Energy Re-use, as further illustrated in (530) of Figure (5-6). Photon Energy Absorption Process (531) of Figure (5-7), of course, can occur in an instant of time.

In Application of Usage

The Hydrogen Fracturing Process (390) of Figure (3-42) simply triggers and releases atomic energy (gtnt) from natural water (85) of Figure (3-26) by retarding and preventing the reformation of the water molecule being subjected to sub-critical state (520) of Figure (5-3) during thermal gas ignition (100) of Figure (4-8) as to (70) of Figure (4-5). The Voltage Intensifier Circuit (110) of Figure (4-9) brings on the "Electrical Polarization Process" (160) of Figure (3-26) that switches off the covalent bond (550) of Figure (5-8) of the water molecule without amp influxing. Energy Pumping Action (500) of Figure (5-1) as to (510) of Figure (5-2) undergoing "Resonant Propagation" of opposite electrical stress (ST-ST'/RU-RU') oscillates the "Energy Aperture" (7a ~~~ 7n) of the combustible gas atoms to increase the atomic energy level (see graph 610 of Figure 6-4) of the combustible gas atoms prior to gas ignition. The Electron Extraction Process (230) of Figure (3-30) as to (270) of (3-34) ionizes the highly energized combustible gases to decrease atomic mass while applied traveling voltage wave-form (57) of Figure (6-2) of opposite electrical polarity (E9-66/ E10-67) initiates the voltage-triggering process (600) of Figure (6-3) once maximum voltage deflection (Vo ~ Va ~ Vb ~ Vc ~ Vn) of Figure (6-4) is achieved at Activation-Point (E9d) of Figure (6-2)...releasing thermal Explosive Energy (gtnt) from the atomic level of the water molecule. The energy contained in a gallon of water exceeds 2.5 million barrels of oil when equated in terms of atomic energy. Water, of course, is free, abundant and energy-recyclable.

Stanley A. Meyer

RE: Atomic Energy Balance of Water   Memo WFC 424

Covalent Switch-Off

Covalent Switch-Off occurs when deflected and elongated Orbital Electron Pathway (541) reaches a point where applied Opposite Electrical Stress (ST-ST'/RU-RU') (A-A'/Z-Z') is sufficient enough to cause the Gyroscopic Action (542) of Nucleus Particles (543a xxx 543n) to be reduced in orbital spin-velocity...which, when occurring, directly weakens the covalent bonding of the water molecule (q-q') by attenuating the electromagnetic fields of each atom structure of the water molecule (210) of Figure (3-27) (Memo WFC 422DA) being subjected to and undergoing Electrical Polarization Process (160) of Figure (3-26) (Memo WFC 422DA), as further illustrated in (550) of Figure (5-8).

Atomic Energy Equilibrium

Force Factor (548) (Particle Mass $\underline{545}$ moving through a Electrostatic Field $\underline{544}$ having a Negative Electrical Charge $\underline{546}$ and exhibiting a Electromagnetic Field $\underline{547}$) opposing Orbital Velocity (549) of moving electron (s) (1a xxx 1n) is/are being continually compensated for and overcome when a discrete amount of Universal Energy (9) enters into and interacts with the Energy Spectrum of the atom by way of energy aperture (7)... allowing and maintaining Stable Energy-State of the atom even if other energy stimuli are not available beyond the physical embodiment of the atom ... such as the absent of sunlight (Photon) absorption. *See Appendix (B) Note (3).*

Energy Aperture of the Atom

Energy Aperture (7) of Figure (5-8) exists in all atomic structures (individual atoms) and functions as a one-way energy valve when the Incoming Energy Vortex transfers a given or discrete amount of "Universal Energy" (having higher energy potential) into the Energy Spectrum of the Atom to compensate for and maintain "Atomic Energy Equilibrium" during either "quiescent" or "active" state of tickling particle oscillation as an energy generator (example, bouncing electrons in a sinusoidal-wave form to its orbital path to cause electromagnetic wave oscillations) ... allowing the influx of "Universal Energy" (Light Energy) to vortex inwardly toward the centre part of the "Gyroscopic" orbital spin-velocity of nucleus particles being displaced about a common axis of rotation ... the resulting "Gyroscopic Action" regulating the inward flow or flow intensity of "Universal Energy" in direct relationship to the orbital spin-velocity of the geometrical particle-structuring (interlocking particles grouped in space relationship to each other) set in orbital motion, as illustrated in (570) of Figure (5-10).

RE: Atomic Energy Balance of Water   Memo WFC 424

Solar Energy Actuator

Energy Recycling Spectrum (530) is achieved since expelling de-energized water molecule (s) (538a xxx 538n) of Figure (5-7) is/are opto-sensitive to photon energy (537a xxx 537n) in that photon-energy (537) is composed of electromagnetic radiant energy transmitted through the medium of space by way of pulse-oscillations known as "frequency of wavelength" and, is expressed by a quantum of Electromagnetic energy in the following equation

$$E = h\nu \qquad (Eq\ 31)$$

Where,

(E) is the energy, (h) is Planck's Constant (energy x time) (6.547×10^{-27} erg-second), and (ν) is the frequency associated with the photon.

The momentum of the photon in the direction of propagation transversing the medium of space is, thus, expressed by the following equation

$$h\nu / c \qquad (Eq\ 32)$$

Where,

(c) is the quantum of electromagnetic energy carried in a small amount and moving with the speed of light, as so illustrated in Figure (5-11). Optical Photon having energies corresponding to wavelengths between 120 and 1800 nanometers ... thus, propagating sun's light or its direct rays (539a xxx 539n) of Figure (5-6).

Exposing the expelling de-energized water droplets (531) of Figure (5-6) to incoming sun's light (534) of Figure (5-7), now, causes Photon Absorption Process (537) to deflect the orbital electron (s) to a higher energy-state (538) away from the atom nucleus as so illustrated in (520) of Figure (5-3) as to (540) of Figure (5-7) once the atom nucleus absorbs the inflowing photon energy (537). The deflected electron (s), in turn, applies an increase/greater electrical-stress (A'A) onto the gyroscopic spin-velocity of the nucleus particles (570) of Figure (5-10) as to (550) of Figure (5-8) ... causing Energy Aperture (7) of Figure (5-10) to oscillate as a "Energy Generator" ... releasing Universal Energy (9) of Figure (5-10) into the atom nucleus ... allowing Atomic Energy Level Adjustment (540) of Figure (5-7) to take place, as illustrated in (970) of Figure (5-12) ... re-energizing the water molecule for hydrogen reuse, as illustrated in (530) of Figure (5-6).

Stanley A. Meyer

RE: Atomic Energy Balance of Water — Memo WFC 424

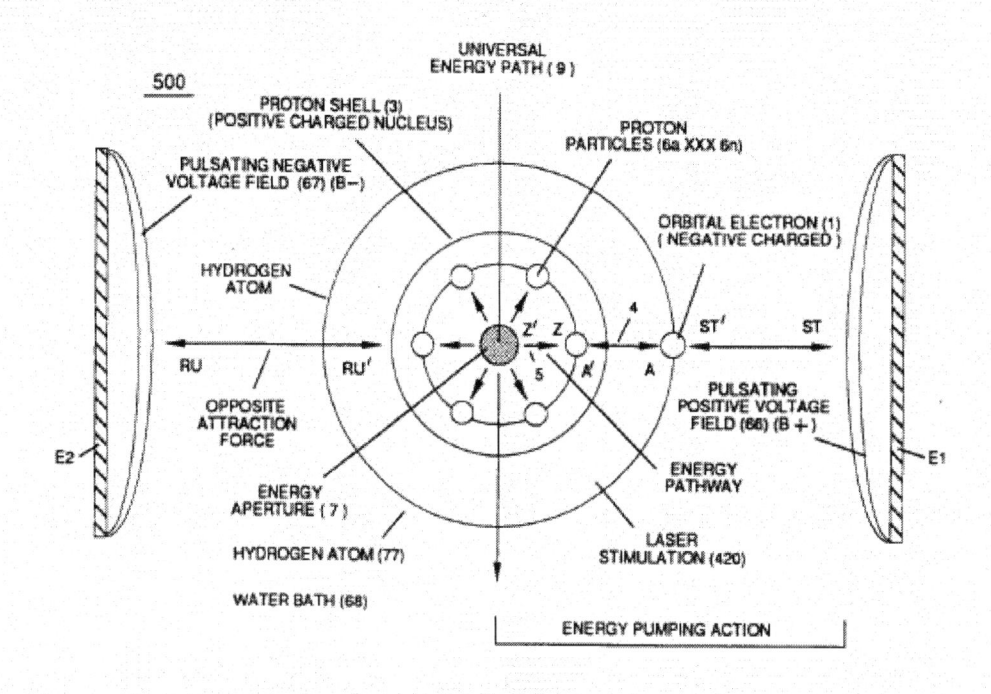

FIGURE 5-1: HYDROGEN ENERGY BALANCE

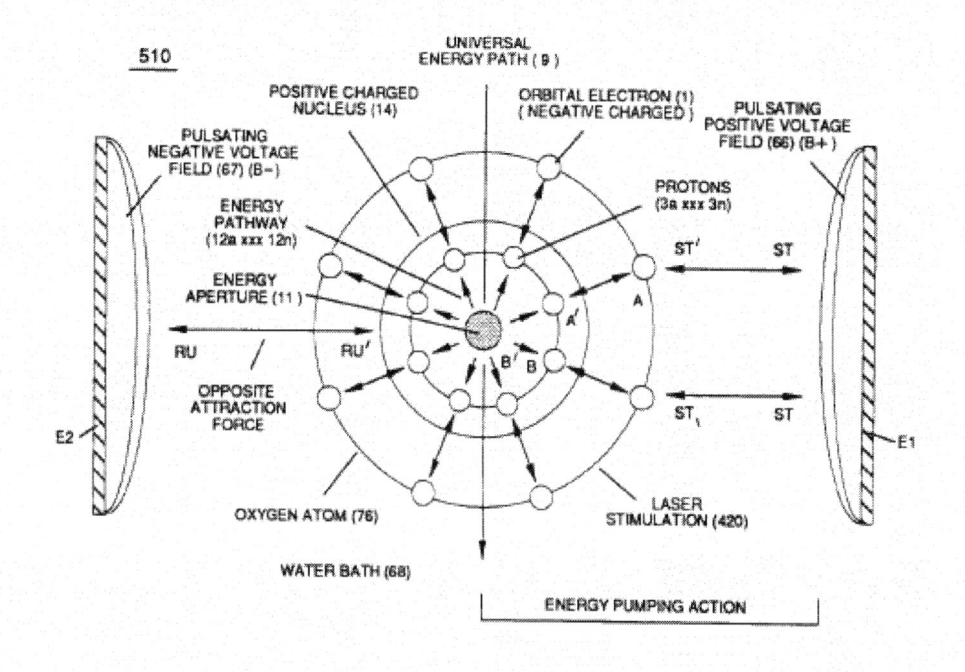

FIGURE 5-2: OXYGEN ENERGY BALANCE

Stanley A. Meyer

RE: Atomic Energy Balance of Water   Memo WFC 424

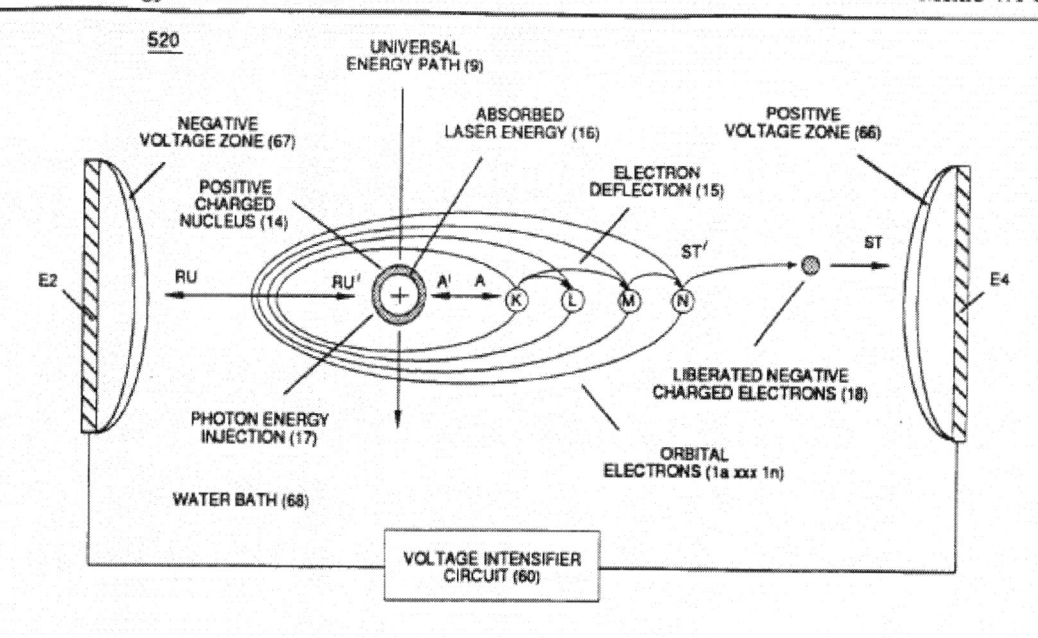

FIGURE 5-3: ENERGY PUMPING ACTION

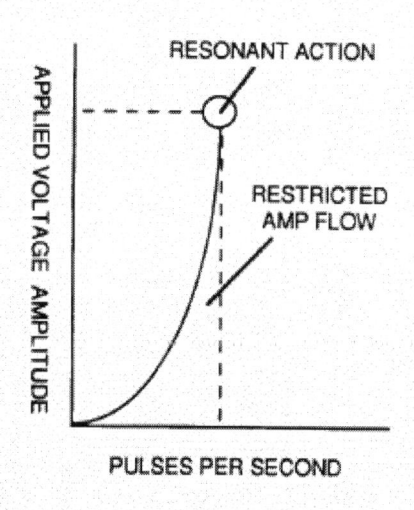

FIG. 5-4A: TRIGGERING RESONANCE

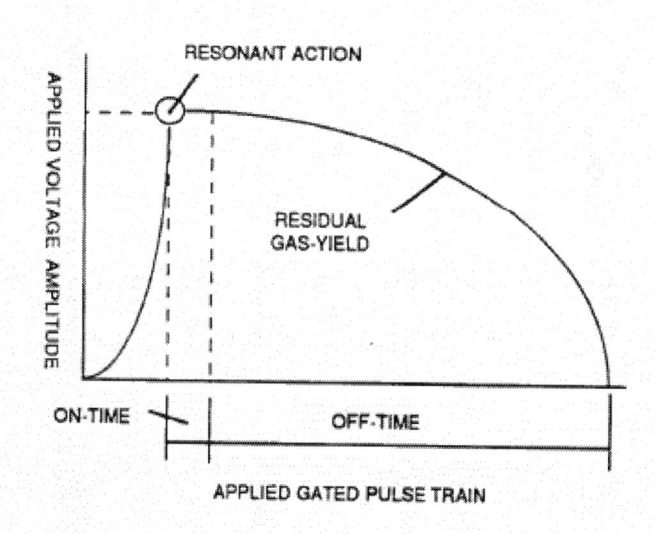

FIG. 5-4B: SUSTAINING RESONANCE

Stanley A. Meyer

5-8

RE: Atomic Energy Balance of Water Memo WFC 424

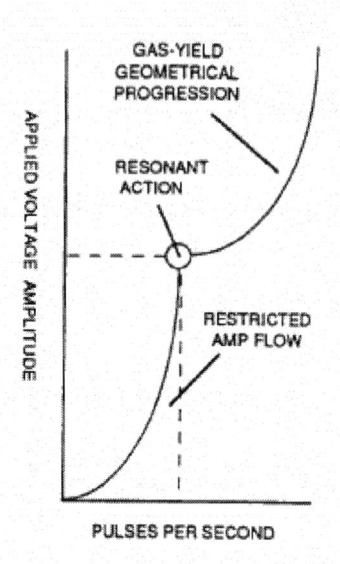

FIGURE 5-4C: RESONANT PROPAGATION

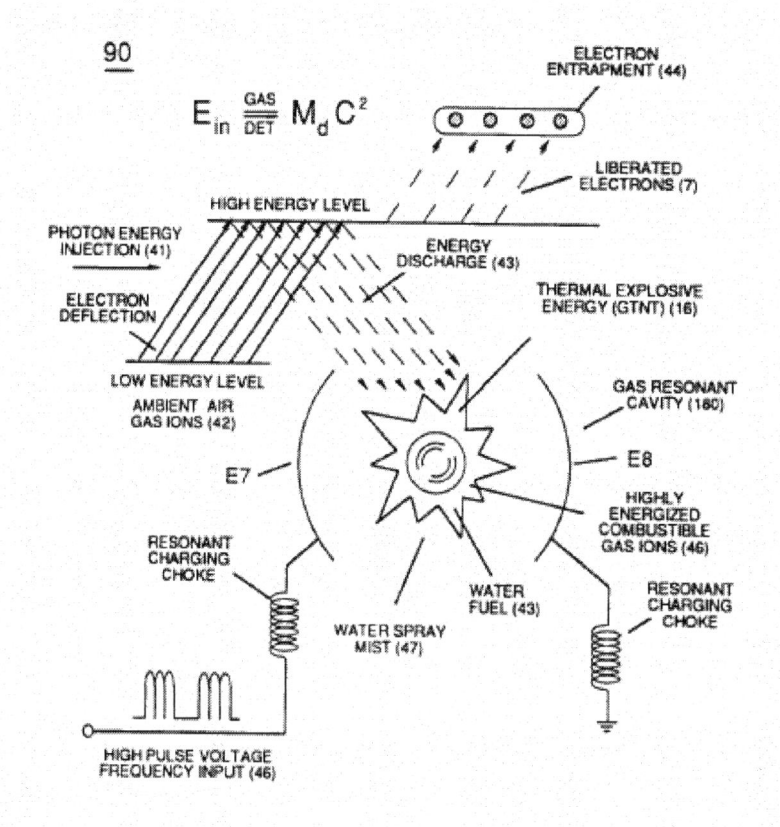

$$E_{in} \underset{DET}{\overset{GAS}{\rightleftharpoons}} M_d C^2$$

FIGURE 5-5: VOLTAGE IGNITION

RE: Atomic Energy Balance of Water Memo WFC 424

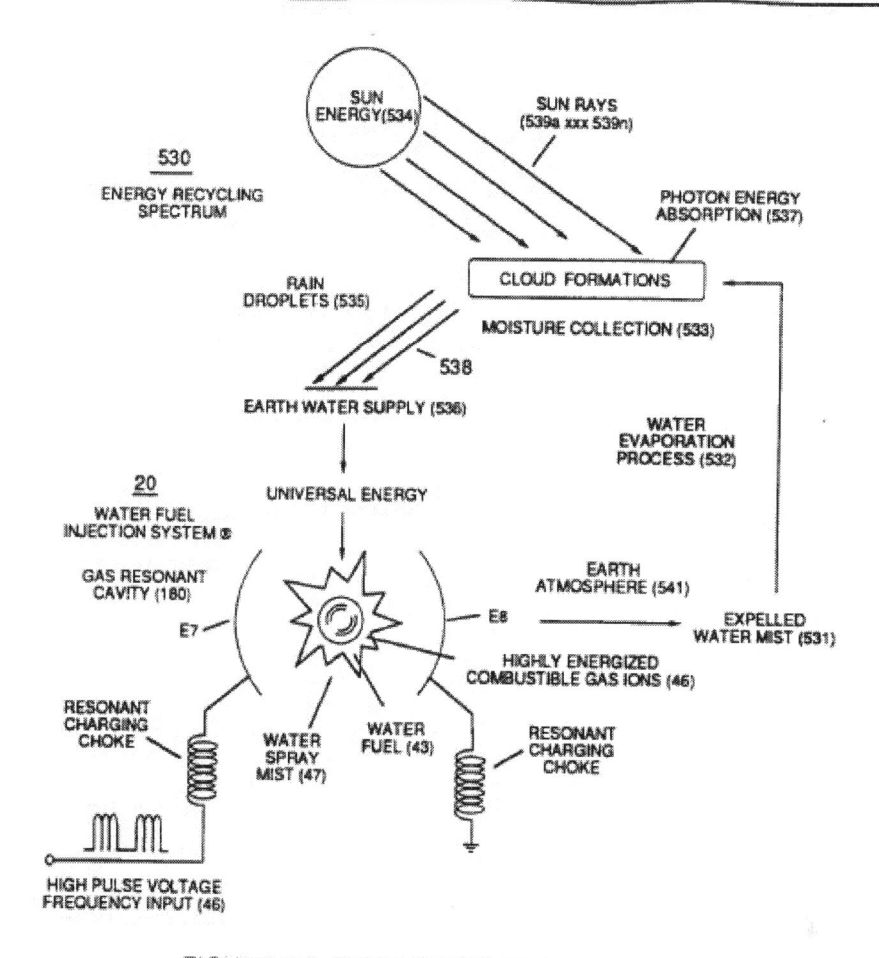

FIGURE 5-6: OPEN ENDED ENERGY SYSTEM

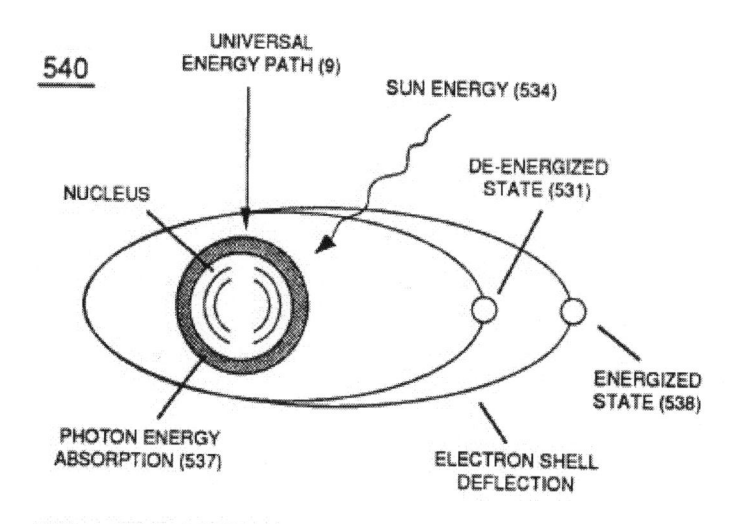

FIGURE 5-7: ATOMIC ENERGY LEVEL ADJUSTMENT

RE: Atomic Energy Balance of Water Memo WFC 424

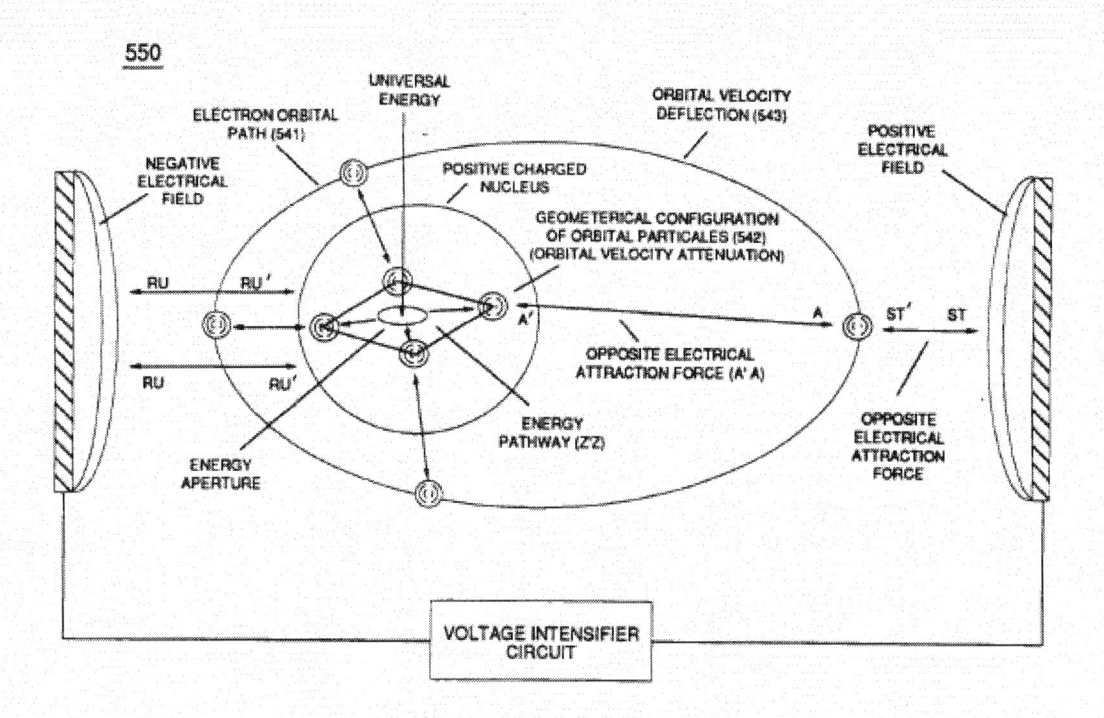

FIGURE 5-8: COVALENT SWITCH-OFF

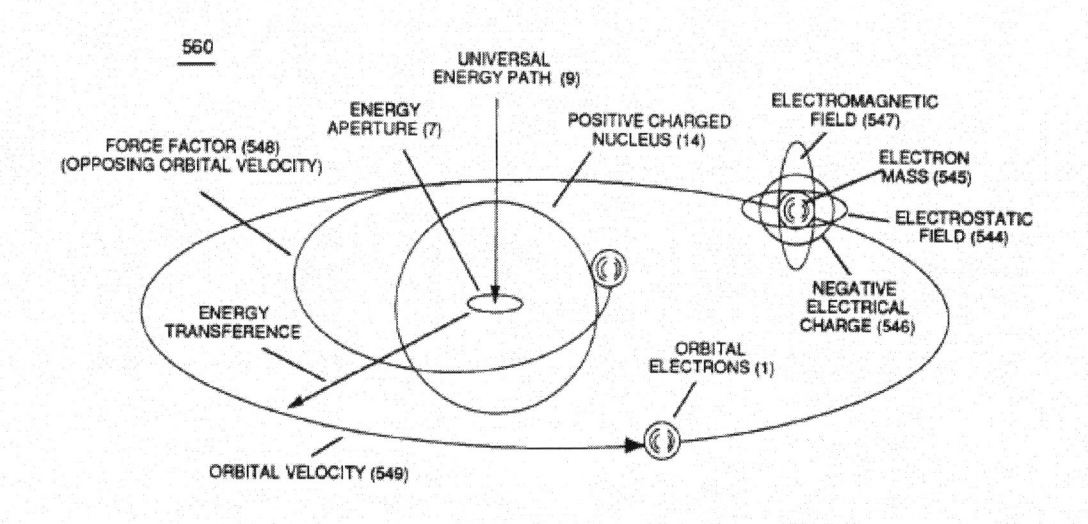

FIGURE 5-9: ATOMIC ENERGY EQUILIBRIUM

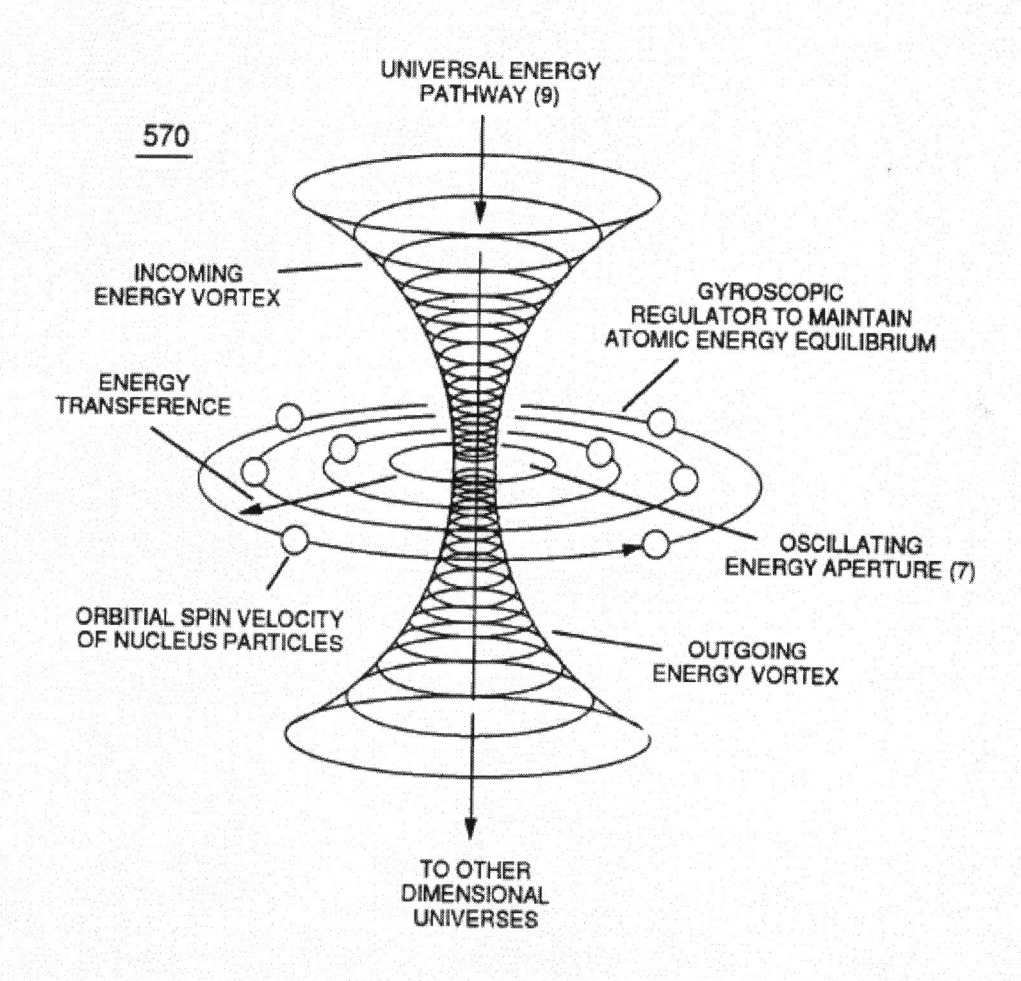

FIGURE 5-10: ENERGY APERTURE OF THE ATOM

RE: Atomic Energy Balance of Water   Memo WFC 424

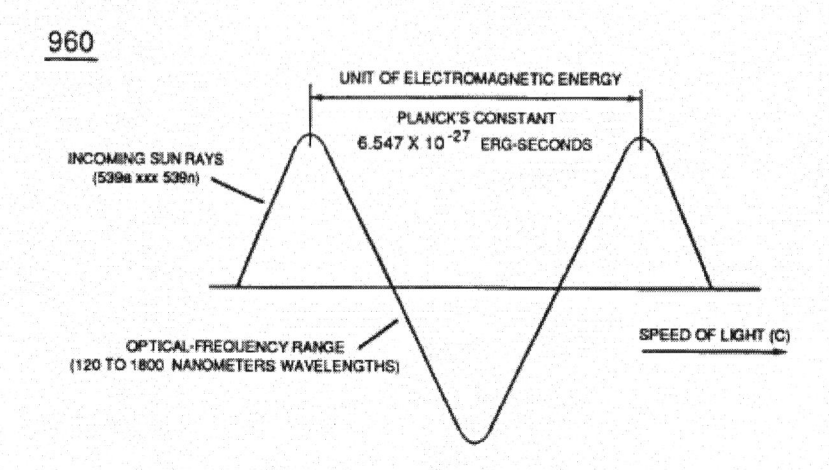

960

UNIT OF ELECTROMAGNETIC ENERGY

PLANCK'S CONSTANT
6.547×10^{-27} ERG-SECONDS

INCOMING SUN RAYS
(539a xxx 539n)

OPTICAL-FREQUENCY RANGE
(120 TO 1800 NANOMETERS WAVELENGTHS)

SPEED OF LIGHT (C) →

FIGURE 5-11: OPTICAL PHOTON

970

OXYGEN ATOM (76)

RE-ENERGIZED WATER MOLECULE (535)

INCOMING PHOTON ENERGY FROM THE SUN (537)

GREATER ELECTRICAL STRESS
(q' ~ qa xxx q' ~ qn)

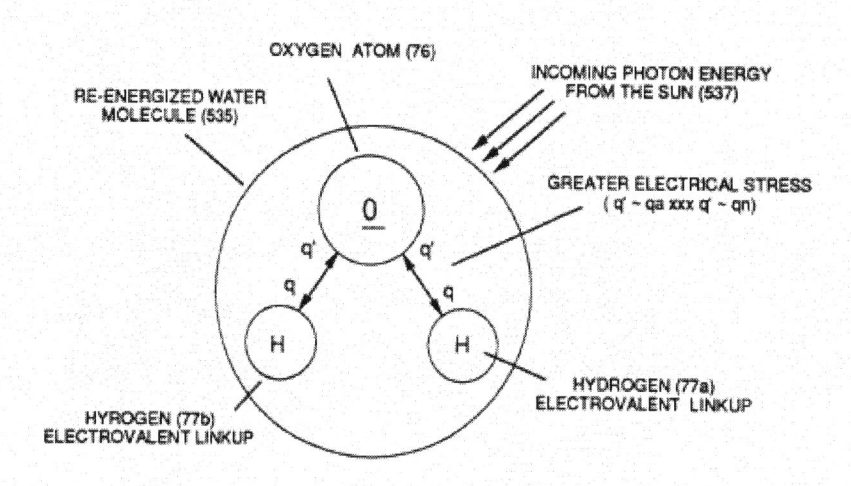

HYDROGEN (77a) ELECTROVALENT LINKUP

HYROGEN (77b) ELECTROVALENT LINKUP

FIGURE 5-12: ENERGY ACTUATOR

Stanley A. Meyer   5-13

Memo WFC 425

Water Fuel Injector: Taper Resonant Cavity

Tuning-in to the dielectric properties of water by way of voltage stimulation allows a sequent of events to occur in a instant of time:

Superimposing "Electrical Stress" of opposite voltage polarity to switch-off covalent bonding to perform Electrical Polarization Process of water.

Propagating the use of Universal Energy by " Energy - Aperture" oscillation of the release combustible gas atoms of water.

Encouraging particle oscillation as an "Energy Generator" to further "Energy Prime" the combustible gas ions prior to gas-ignition once electron ejection occurs.

Triggering the highly "energized" combustible gas atoms of water by way of "atomic electrical discharged" as an ever increasing electrostatic pressure of opposite voltage polarity under pulse resonance is applied across the water molecule ... releasing thermal explosive energy (gtnt) on demand in the following way:

Section 6

RE: Water Fuel Injection System　　　　　　　　　　　　　　　　Memo WFC 425

Water Fuel Injector

(Taper Resonant Cavity)

Voltage potential of opposite electrical polarity (ST - ST' ~ RU - RU') of Figure (5-1) (Memo WFC 424) titled "Atomic Energy Balance of Water" is further enhanced by simply electrically interfacing voltage intensifier (VIC) circuit coil-assembly (580) of Figure (6-1) with "Taper Resonant Cavity" (590) of Figure (6-2), as schematically illustrated in (60) of Figure (3-22) as to pulse core configuration (190) of Figure (3-23) (Memo WFC 422 DA) titled "WFC Hydrogen Gas Management System.

As incoming gated pulse-train (46a xxx 46n) of Figure (3-17) is electronically "tuned" to adjust pulse off-time (T2) to compensate for "rise" and "fall" of magnetic field coupling (71a xxx 71n) for a predetermined resonant pulse-frequency established and determined by the dielectric value of natural water in direct relationship to resonant cavity geometrical configuration...dielectric value of water being 78.54 since water molecule (85) oxygen atom "L" orbit (76) occupies the maximum allowance eight electrons (79a xxx 79n), calibrated gated unipolar pulse train (64a xxx 64n) of Figure (3-20) is outputted from resonant choke (56) and electrically transmitted to positive outer conical surface (E9); while, at the same time, negative potential of electrical intensity of force (67) (negative voltage potential) is electrically directed to inner conical surface (E10), forming an "open-air" conical cavity (570) having parallel sides (in other cases non-linear voltage-surfaces) in space relationship (typically .010 gap) with diminishing circumference-area (E9a xxx E9n) / E10a xxx E10n) in linear progression. Together, parallel sides (E9 / E10) not only functions as a "voltage wave-guide" (570) but, also, acts and performs as a "voltage intensifier circuit" when applied gated pulse-frequency (64a xxx 64n) travels the length of conical cavity (570) toward exit port (32). At each progressive point of diminishing circumference surface-area (E9a ~ b ~ c ~ d ~ E9n) voltage amplitude intensity increases (Vna ~ b ~ c ~ d ~ Vnn) uniformly, as illustrated in (600) of Figure (6-3) as to Traveling Voltage Wave-forms (730a ~ b ~ c) of Figure (7-12), see WFC Memo (426).

Activation point (E9a) exposes water flow (85) to voltage wave-form (64) of Figure (6-1) to begin water-to-energy conversion process (100); at activation point (E9b) voltage intensity is

Stanley A. Meyer

RE: Water Fuel Injection System   Memo WFC 425
___

increased sufficiently to perform Electrical Polarization Process (160) of Figure (3-26); onward toward activation point (E9c) and beyond universal energy priming stage (500) of Figure (5-1) occurs; once activation point (E9c) is reached Gas Ionization Process (230) of Figure (3-30) takes place; and finally, activation point (E9d) thermally ignites (atomic agitation) the "Energy-Primed" combustible gas-mixture (520) of Figure (5-3) as to (100) of Figure (4-8) by "electrostatic discharge" while being subjected to ever increasing "electrostatic pressure". All activation points (E9a ~ b ~ c ~ d) performing their respective functions in sequential order in an instant of time since applied voltage level of intensity (typically 20,000 input volts or so) can be extended or increased up to and beyond 90,000 volts range within a millisecond or less.

Taper Water Fuel Injectors

Voltage wave-guide (570) allows the activation points (E9a xxx E9n) to transpire since wave-guide (570), now, functions as a Quenching Circuit (370) of Figure (3-40) to prevent gas ignition until the traveling gases (under static pressure) are exited out of and away from exit port (32) of Figure (6-2) ... producing thermal explosive energy-yield (16), as further illustrated in (70) of Figure (4-5) titled "Voltage Triggering". Basically, then, activation process (590), now, design - forms Water Fuel Injector (20) of Figure (4-2) as to (130) of Figure (4-11) (Memo WFC 423 DA) ... allowing Water Fuel Injectors (20a xxx 20n) to replace standard Internal Combustion Engine spark-plugs and fossil-fuel injector ports, as graphically illustrated in Figure (140) of Figure (4-12) titled "Furnace Retrofit," (150) of Figure (4-13) titled "Jet Engine retrofit," and (160) of Figure (4-14) titled "Rocket Engines Retrofit."

Voltage Intensifier Coil-Assembly

Activation Process (590) of Figure (6-2) as to (100) of Figure (4-8) is achieved since amp flow is restricted to enter into Voltage Triggering Process (70) of Figure (4-5) by way of voltage intensifier coil-assembly (580) of Figure (6-1). Inherently, the design parameters of coil-structures (580) of Figure (6-1) determines "Efficiency" (minimizing amp leakage) by which "Voltage Intensity of Opposite Potential" (600) of Figure (6-3) can perform work to trigger Hydrogen Fracturing Process (520) of Figure (5-3) (Memo WFC 424 DA) as to (100) of Figure (4-8) (Memo WFC 423DA), as graphically denoted in (750) of Figure (7-14) of WFC memo (425) titled VIC Matrix Circuit - Instant Explosion of Water.

RE: Water Fuel Injection System   Memo WFC 425

Tri - Coil Construction

    Resonant Choke Coils (56 / 62) of Figure (3-23) (Memo WFC 422 DA) are composed of 430F or 430FR inductance stainless steel film coated (hi dielectric value) wire (typically .004 Ga. or smaller) which are axially (spiraled) Bifilar wound about core bobbin (502), forming individual spiral-wrap (inner to outer circumference and being equally-length) coils (501a xxx 501n) electrically connected in sequential order to form resistive pickup coil (503).

    Primary Coil (26) (typically .030 Ga.) film coated magnet wire is longitudinal wrapped in space relationship on top of and layered bidirectional (507a xxx 507n) across spiral-wrap coils (501a xxx 501n) to complete bobbin cavity (504).

    Secondary pickup coil (52) of Figure (3-23) is, also, composed of individual spiral-wrapped coils (505a xxx 505n) (typically .002 Ga. magnet wire) electrically connected in sequential order to form bobbin cavity (506) which is placed on top of and in space relationship to primary coil cavity (504).

    Resonant bobbin assembly (503), primary bobbin assembly (504), and secondary bobbin assembly (506), now, make up and structurally forms voltage intensifier (VIC) coil-assembly (530) of Figure (4-6) when electrical steel core material (53) forms a close-loop magnetic induction pathway centrally through and around (VIC) coil-assembly (530), as schematically illustrated in (190) of Figure (3-23) (Memo WFC 422 DA).

Electromagnetic Interaction

    The resultant tri-coil configuration (Inductance core __53__ - choke coils __56/62__ - primary coil __26__ - secondary coil __52__), now, allows magnetic field coupling (71a xxx 71n) to pass through both resonant-coils (56/62) and secondary coil (52) simultaneously when primary coil (26) is pulsed energized by way of incoming pulse-train (46a xxx 46n). In doing so, magnetic flux-lines (71a xxx 71n) are induced into spiral-wrap coils (505a xxx 505n) to produce inductance coupling (511a xxx 511n) between each secondary spiral-coils (505a xxx 505n) which are parallel formed to expanding magnetic flux-lines (71a xxx 71n) ...producing step up voltage potential of positive electrical intensity (positive voltage potential) by way of inductance / capacitance interaction across secondary coil-assembly (52) while keeping opposition to electromagnetic build up to a minimum.

Magnetic flux-lines (71a xxx 71n) being emitted on the opposite side of primary coil (26) induces further increase in positive voltage potential (64 a xxx) since inductance / capacitance (Cd / DL) of Figure (7-3) interaction is, also, occurring in both resonant charging chokes simultaneously in balance relationship to the same pulse sequent (46)... producing inductance coupling (512a xxx 512n) (Rp /Rp1/Rp2) of Figure (7-8) in parallel relationship to expanding field (71), as before. The resultant Pulsing Sequence (49a xxx ~ T1/T2 ~ xxx 49n) of Figure (7-1) allows voltage (T1) across Inductance Chokes (56/62) while current flow lags by 90°.

Together, external magnetic field (71), inductance coupling field (512a xxx 512n), resistive value ($Z_2 + Z_3$) of stainless steel wire-coil (56/62), and the dielectric value (ohmic or resistive value) (R_e) of water aids and performs amp restriction process (520) of Figure (5-3) while allowing applied voltage amplitude to be electrically transmitted without signal degradation. (see circuit resistive equations (Eq 9) (Memo WFC 420, once again) as to VIC Matrix Circuit (690) of Figure (7-8).

The resultant dynamic voltage potential (600) of Figure (6-3), now, performs the Hydrogen Fracturing Process (390) of Figure (3-42) (Memo WFC 422 DA) in such a way as to allow particle oscillation to take place as a "Energy Generator" (see Memo WFC 424 titled Atomic Energy Balance of Water) to further enhance thermal explosive energy-yield (16a xxx 16n) (70) of Figure (4-5) (Memo WFC 423 DA), as graphically illustrated in (610) of Figure (6-4).

Injector (590) of Figure (6-2) and voltage intensifier coil-circuit (580) of Figure (6-1) as to (190) of Figure (3-23) is electronically Interlinked with Water Fuel Management (WFMS) System (40) of Figure (4-2) (Memo WFC 423DA) to form "The Water Fuel Injection System" ® that triggers and performs voltage activation process (600) of Figure (6-3) as to (730) of Figure (7-12).

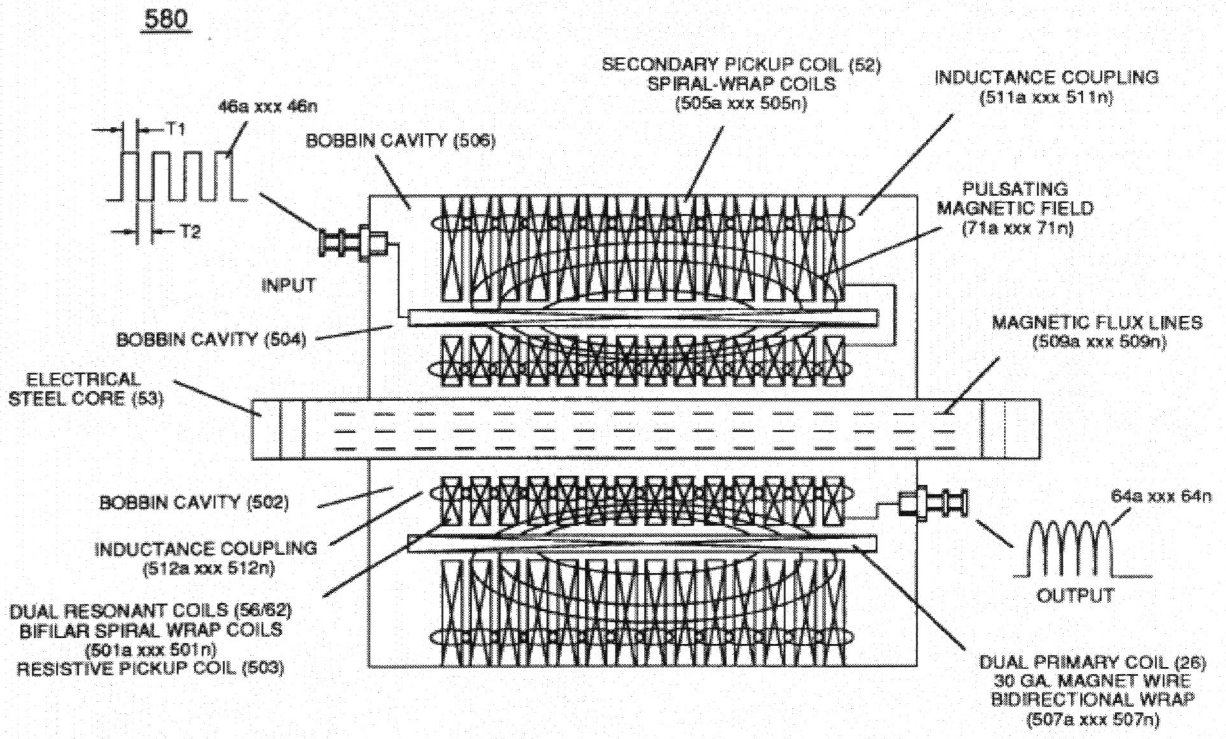

FIGURE 6-1: (VIC) COIL ASSEMBLY

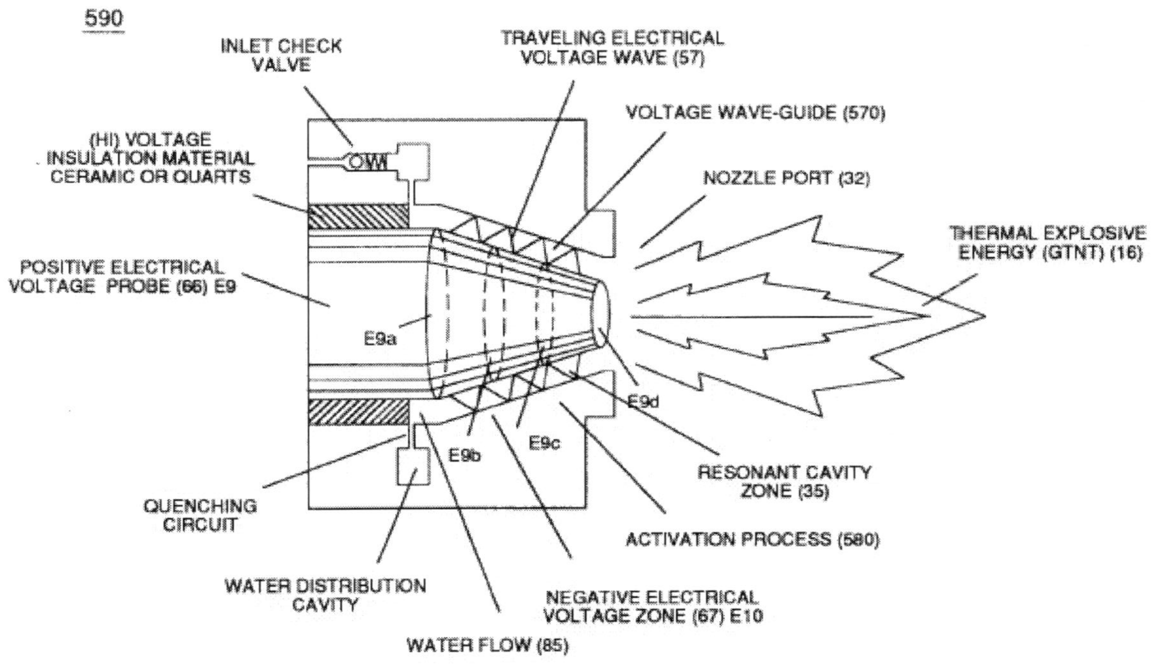

FIGURE 6-2: TAPER RESONANT CAVITY

RE: WATER FUEL INJECTION SYSTEM

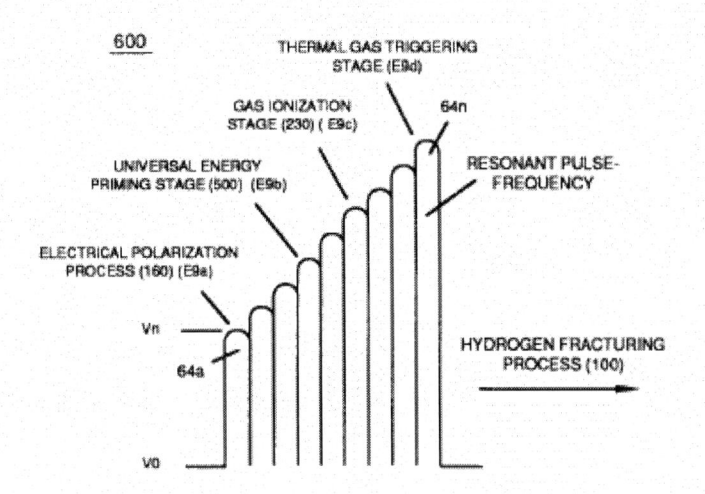

FIGURE 6-3: DYNAMIC VOLTAGE POTENTIAL

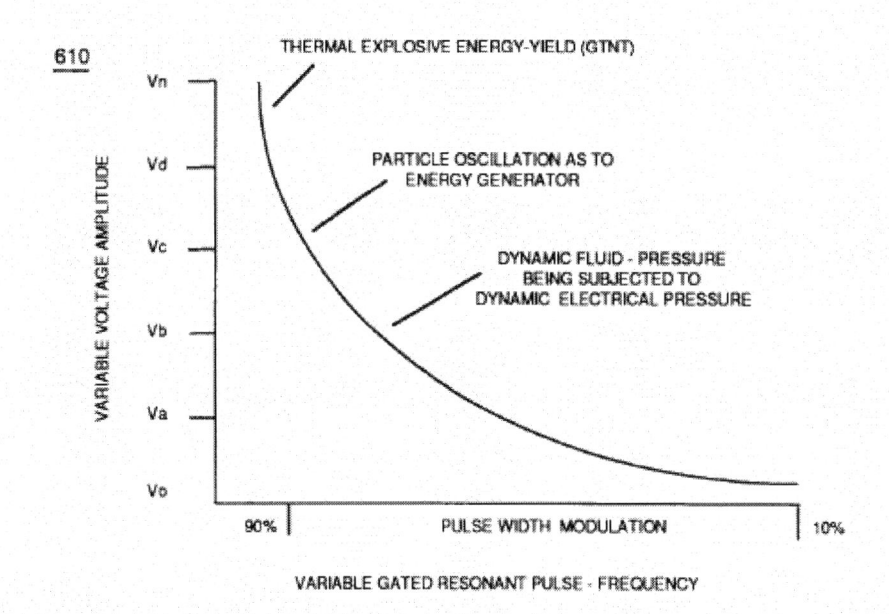

FIGURE 6-4: PARTICLE OSCILLATION AS TO ENERGY GENERATOR

Memo WFC 426

VIC Matrix Circuit

Instant Explosion of Water

VIC Coil Assembly is specially designed to allow Voltage Potential of "opposite electrical attraction force" of High Voltage Intensity" to " instantly" release Thermal Explosive Energy (gtnt) from natural water.

The Voltage Intensifier Circuit takes advantage of the "Electron Bounce Phenomenon" to trigger Hydrogen Fracturing Process without amp influxing.

Interlinked with VIC Coil Assembly, the Water Fuel Injector acts and performs as a "Voltage Amplifier" by simply altering the Voltage Wave Guide to either form a "compressional" or "Expanded" Voltage Wave Form that increases Electrical Voltage intensity beyond applied excitation voltage outputted from VIC Coil.

The "Mode of Operability" of VIC Coil Assembly is systematically activated by a programmable signal input, and is performed in the following way:

Section 7

RE: VIC Matrix Circuit   Memo WFC 426

VIC Matrix Circuit

Instant Explosion of Water

The "mode-of-operability" of VIC Coil Assembly allows Voltage Potential of opposite voltage polarity to increase and be attenuated up to and beyond 20 Kilovolts while inhibiting and restricting amp leakage in the milliamperes range ... establishing operational parameter of utilizing "Opposite Electrical Attraction Force" of "high voltage intensity" to "instantly" releases thermal explosive energy (gtnt) from natural water. The Voltage Intensifier Circuit takes advantage of the "Electron Bounce Phenomenon" to trigger Hydrogen Fracturing Process without amp influxing. Taper Resonant Cavity functions as a "Voltage Amplifier" when interlinked with VIC Circuit.

Voltage Intensifier Circuit (60) of Figure (3-22) (Memo WFC 422 DA) as to Figure (1-1) (Memo WFC 420) and Voltage Intensifier Circuit (620) of Figure (7-1) are specifically designed to restrict amp flow during Programmable Pulsing Operations (49a xxx 49n) but in different operational modes: VIC voltage circuit (60) utilizes copper wire-wrap to form Resonant Charging Chokes (56/62) of Figure (3-22) in conjunction with Switching Diode (55) to encourage and make use of "Electron Bounce" phenomena (700) of Figure (7-9) to help promote Step Charging Effect (628) of Figure (7-7) by preventing electrical discharge of Resonant Cavity (140 ~ 170) since Blocking Diode functions as an "Open" switch during Pulse Off-time; whereas, VIC Voltage Enhancement Circuit (VIC - VE) (620) of Figure (7-1) incorporates the use of stainless steel wire-wrap coils (614/615) to accomplish the formation of unipolar gated pulse-wave (64a xxx T3 xxx 64n) without experiencing "signal distortion" or "signal degradation" (preventing transformer ringing during signal propagation) as elevated voltage levels (~ xx Vc ~ xx Vd ~ xx Vn) while allowing the reduction of Capacitor-Gap (Cp) (616) of Figure (7-11) width spacing (57 of Figure 3-25 ~ 35 of Figure 6-2) (typically .060 ~ .010) respectively, as illustrated in Tubular Resonant Cavity (170) as to Taper Resonant Cavity (620) of Figure (7-1).

Switching Diode (55) of Figure (3-22) prevents Bidirectional electron flow (current flow in one direction only) since Blocking - Diode (55) only conducts "current flow" in the direction of schematic-arrow while being placed in-line with VIC Circuit impedance interaction (R1 + Z2 + Z3 + Re), as mathematically extrapolated in Circuit Equation (Eq 9) ... Diode (55) being placed between Secondary Pickup Coil (52) and Resonant Charging Choke (56) to act as an electronic switch in open-position during pulse off-time (T2) of Figure (7-8) while preventing electron flow

Stanley A. Meyer   7 - 1

RE: VIC Matrix Circuit   Memo WFC 426

in reverse direction when Inductor (L1) collapsing electromagnetic field (FL1) produces another unipolar pulse wave-form (64a ~ 64b) ... producing unipolar voltage wave-form (64a xxx 64n) during repeated pulse-signal (46a xxx 46n) on-time (T1a xxx T1n) ... allowing the formation of an gated pulse- frequency pulse-train (64a/64b ~ T3 ~ 64a/64b) when pulse off-time (T3) is greater than time-period (T2) ... input-signal (49a xxx 49n) being a Pulse-Train where (T2) pulse off-time (T2) is adjusted to allows Unipolar Pulse-Train (64a xxx T3 xxx 64n) ... outputting Voltage-wave signal (64a xxx 64n) being a pulse-frequency doubler due to Inductance Reactance (FL) of Inductor Coil (56) of Figure (3-22) when collapsing magnetic field (Fl) of Figure (7-3b) re-cuts coil-wrap (L1) during each pulse off-time (T2) ... producing a second unipolar voltage wave-form (64b) during the rise and fall of magnetic field (71), as further illustrated in (620) of Figure (7-1).

Resistance (Rs)

In reference to the use of stainless steel (s/s) coil-wrap (614/615), resistive wire value (Rs1/Rs2) of Figure (7-8) (typically 11.6 K ohms per coil) is sufficient enough to inhibit current flow oscillation in direct relationship to circuit impedance (Eq. 9) since "current flow" is, also, restricted in the milliampere (s) range due to (s/s) wire material (Rs1/Rs2) composition ability to oppose electron interaction or electron interchange from one atomic structure to another; while, at the same time, conducting and permitting the transmission of "Voltage Potential" across circumference surface area (skin effect) (66/67) of Figure (7-11) as to Figure (590) of Figure (6-2) to bring-on and perform Voltage Wave-Guide phenomena (57) of Figure (6-2) ... causing and allowing the applied Electrical Stress of opposite voltage polarity (ST-ST' ~ RU-RU') to trigger Hydrogen Fracturing Process (390) of Figure (3-42) in an instant of time ... releasing thermal explosive energy (gtnt) (16) of Figure (4-5) on demand from natural water (85) of Figure (3-26) since the dielectric value (Re) of (Eq.9) of Water Fuel (85) is further approximated in Capacitance Equation (Eq.22), as illustrated in (650) of Figure (7-4) as to Tapered Voltage Wave-Guide (720) of Figure (7-11) ... allowing the dielectric value of Water (Re) to be a part of Voltage Intensifier Circuit (110) of Figure (4-9) capability of restricting amp flow during Voltage Pulsing Operation (49a xxx 49n) of (620) of Figure (7-1) as to VIC Matrix Circuit (690) of Figure (7-8) ... allowing applied opposite Voltage potential (ST-ST' ~ RU-RU') to perform work without amp "influxing", as systematically depicted in VIC Matrix Circuit (690) below.

Inductance (FL)

Component Interaction promotes Component Reactance during D.C. pulsing operations

Stanley A. Meyer

RE: VIC Matrix Circuit   Memo WFC 426

while allowing variable voltage amplitude (Vo ~ Va ~Vb ~ Vn) of Figure (7-13) to be attenuated independently of Voltage Pulse frequency (49a xxx 49n), as so illustrated in (600) of Figure (6-3).

Resonant Charging Circuit (630) of Figure (7-2) being an LC Circuit is formed when Inductor (614) of Figure (7-1) is electrically linked to Taper Capacitor (720) of Figure (7-11) in series arrangement. Inductor (614) is an insulated wire wound in a spiral pathway around Bobbin Cavity (580) of Figure (6-1) to form Voltage Stepping Coil (710) of Figure (7-10) as to (580) of Figure (6-1). Capacitor (E9/E10) of figure (6-2) as to (720) of Figure (7-11) is formed when outer tapered surface (66) and inner tapered surface (67) forms Water-Gap (616) of Figure (7-11) as to Figure (590) of Figure (6-2) having placed there between Dielectric Water Bath (85/Re), as schematically illustrated in matrix outline in (670) of Figure (7-6) as to (690) of Figure (7-8) and further detailed in Electrical Charging Effect (650) of Figure (7-4).

Component Reactance to D.C. pulsing transforms inductor (614) of Figure (7-1) / Capacitor (E9/E10) of Figure (7-11) LC circuit of Figure (7-2) into an Resonant Charging Choke (614) which steps up an unipolar oscillation of an given charging frequency with the effective capacitance of an pulse-forming network (64a xxx 64n) of Figure (7-1) as to (600) of Figure (6-3) in order to charge Voltage Zones (E9/E10) to an higher potential beyond applied voltage input ... interacting Distributed Capacitance (Cda xxx Cdn) and Distributed Inductance (DIa xxx DIn) of Figure (7-3) of Inductor Coil (614) of (7-1) with "Electrical Charging Effect" brought on by the dielectric value of water bath (85/Re), as pictorially illustrated in (650) of Figure (7-4). The established Dielectric Value of Water (85) being 78.54 ohms since the electron "L" orbit of the water molecule (210) of Figure (3-27) occupies the maximum allowable number of eight electrons when covalent linkup of unlike oxygen atom (76) and hydrogen atoms (77a/b) occurs ... stabilizing Water molecule (85) into existence ... thereby, maintaining molecular stability of water by opposing the exchange of electrons from an external electron source (amp inducing circuit) beyond molecular structure (85). Electron interaction (movement of electrons through the liquid medium of water) is further inhibited since natural water contaminates (144a xxx 144n) of Figure (3-24) is normally less than 20 ppm ... distilled water, of course, is generally lab-tested 1ppm or less, as illustrated in (760) of Figure (7-15) as to (750) of Figure (7-14.

Capacitance (Cd)

Capacitor (E9/E10) of Figure (7-1) as to Figure (650) of Figure (7-4) in direct relationship to Water Gap (616) becomes Taper Resonant Cavity (720) of Figure (7-11) as shown in (590) of

Stanley A. Meyer   7 - 3

RE: VIC Matrix Circuit   **Memo WFC 426**

Figure (6-2) since Water Gap (616) is occupied by a dielectric liquid (Re) as herein before identified as natural water (85) having no electrolyte added thereto ... generally rain water (85f) (750) of Figure (7-14) being almost free of contaminates due to Water Evaporation Process (532) of Figure (5-6) ... rain water (85f) being an liquid-insulator that restricts the flow of amps ... a resistive liquid (having an ohmic value of 78.54 ohms) that takes on an "Electrical Charge" when applied Voltage Potential (66/67) of Figure (7-1) as to (650) of Figure (7-4) causes and sets up Molecular Polarization Alinement (617) of Figure (7-4) by way of electrical molecular rotation (opposite electrical attraction force to rotate and position particle alinement) of each water Molecule (85a ~ 85b ~ 85c ~ 85n) being subjected to opposite electrical attraction forces (SS' ~ RR'). In like manner, the stainless steel (s/s) T304 material that forms Voltage Zones (E9/E10) undergo particlealinement of its atomic structure within the atomic infrastructure of plate-material (E9/E10) when exposed to the same applied electrical voltage fields (66/67) after a pre-set time ... causing molecular electrical movement to occur within the surface-material (E9/E10) ... which, after occurring, the newly formed molecular electrical orientation (625a xxx 625n) of Figure (7-4) remains in electrical atomic alinement after pulse off-time (T2) ... aiding the transference of voltage potential during pulse on-time (T1) ... allowing the resultant Surface Polarity Effect (skin effect) (624) of Figure (7-7) to supply a sufficient residual atomic "Electrical Charge Field" to help maintain molecular alinement of water atoms (617) during pulsing operations, as illustrated in (680) of Figure (7-7). Inherently, then, Resonant Cavity (720) of Figure (7-11) as to (650) of Figure (7-4) forms capacitor (ER) of Figure (7-1) when the dielectric liquid of water (85) is placed or injected between electrical conducting plates (E9/E10) while applied voltage Potential of opposite polarity (66/67) is directly exposed to Water Molecules (85a xxx 85n), as depicted in Taper Resonant Cavity (590) of Figure (6-2) as to (650) of Figure (7-4).

     Inductor (614) and Inductor (615) of Figure (7-1) as to (670) of Figure (7-6) is wound or coil-wrapped (see multi-layer equation Eq. 20) in such a manner as to increase the magnetic flux intensity (DIa xxx DIn) of Figure (7-3) as to (580) as to Figure (6-1) in reference to (710) of Figure (7-10) between the turns (618a xxx 618n) of coil-wrap (640). The circular-spiral turns of wire (forming parallel electrical surfaces) is separated by an Insulated Dielectric Coating Material which forms a series of capacitors (Cda xxx Cdn) when magnetic flux-lines (619a xxx 619n) produces Electromagnetic Coupling Field (621) during pulse on-time (T1), as illustrated in (640) of Figure (7-3) as to (690) of Figure (7-8). The series resistance value (Rs) in (670) of Figure (7-6) as to (690) of Figure (7-8) and (670) of Figure (7-6) is determined by the composition of the wire material in terms of its ohmic value (electrical resistivity) per given length and diameter cross-section : Resonant Charging Chokes (614/615) 430F/FR 36 AWG (.006) stainless steel (s/s) wire

Stanley A. Meyer   7-4

RE: VIC Matrix Circuit   Memo WFC 426

equals 60 micro ohms per centimeter; Primary Coil (622) 22 AWG (.028) copper wire equals 5.1933 ohms per pound weight ; Secondary Pickup Coil (623) 35 AWG (.007) copper wire equals 1.3K ohms per pound weight. "Pyre-ML" trade name "Himol" polymer coating-material is used to impart thermal and mechanical resistance to the stainless steel (s/s) wire (614/615) coating; both magnet wire sizes (622/623) uses solderable Nysol (Polyurethane Nylon Jacket) insulation enamel coating as a electrical shield-material ... all dielectric coatings having an effective 3KV per mil dielectric value and formulated specifically to endure automotive temperature range from - 40 ° to 155° C.

Inductance Reactance (Rs ~ Cd ~ FL)

Inductance Reactance occurs when resistance (Rs), capacitance (Cd), and Inductance (FL) interacts together during D.C. Pulsing (49a xxx 49n), as schematically depicted in (690) of Figure (7-8).

Inductance Reactance not only increases voltage across water-capacitor (ER) beyond applied Voltage Potential (626) of Figure (7-7) but, also, establishes "Impedance Field" (FL) across Inductors (L1~L2) of Figure (7-6) which acts and performs as Resonant Charging Chokes (614/615) of Figure (7-1) once placed on opposite side of capacitor (ER) forming Resonant Voltage Effect Circuit (670) of Figure (7-6), as illustrated in (620) of Figure (7-1) as to (690) of Figure (7-8). Both Inductors (L1/L2) are Bifilar wound in equal length to optimize the electromagnetic field strength (FL) in equal electromagnetic intensity (FL1 = FL2) to encourage and promote "Electron Bounce" phenomenon (700) of Figure (7-9) while adjusting (programmable pulse wave-form) input signal Pulse-Frequency (49a xx 49n) to "tune-in" to the "dielectric property" (Re) of water (85) ... causing amp flow to be reduce to a minimum value while allowing voltage potential (627) of Figure (7-7) to go toward infinity if the electronic components would allow it to happen, as graphically illustrated in (750) of Figure (7-14). Inductance Field (L1-FL1) performs "Capacitance Charging Effect" (628); while, at the same time, Inductor Field (L2- FL2) restricts electron movement through VIC Impedance Network Circuit (620) of Figure (7-1) since Inductance Field (FL2) locks onto Electrons Magnetic Field (647) of Figure (5-9) to block the movement of electron flow toward Positive Voltage Potential (66) ... thereby preventing and inhibiting electron-flow to pass through or arc-over capacitor water-gap (Cp) of Figure (7-8) ... such electron blocking action is herein called "Electron Inhibiting Effect" (631), as denoted in (670) of Figure (7-6) as to (750) of Figure (7-14). At elevated or higher amplitude voltage levels xxx Ve xxx Vf xxx Vn), primary electromagnetic coupling field (Rp) of Figure (7-8) transmitted by

RE: VIC Matrix Circuit   Memo WFC 426

way of Inductance Pulsing-Core (190) of Figure (3-23) as to VIC Coil Assembly (580) of Figure (6-1) enters into and passes through both Inductors (L1/L2) simultaneously and offers not only further electron-flow restriction (Rp1/Rp2) to both Inductor Chokes (56/62) but automatically increases voltage potential (xxx Vg xxx Vh xxx Vn) of opposite voltage intensity of equal magnitude (66/67) across Resonant Cavity (140 ~170) ... overcoming any potential loss of pulse-signal due to resistive interaction (Rs1/Rs2) of either or both Inductor Cores (L1/L2) wire-material to the formation of Inductance Fields (FL1/FL2) during reoccurring pulse on-time (T1 a xxx T1n). Electron Inhibiting Effect (631) in direct relationship to Voltage Enhancement Effect (528) is accomplished since stainless steel 430F/FR wire-material is "Electromagnetic Inductive" to incoming electromagnetic flux-lines (71a xxx 71n) (Rp) without (s/s) inductor-wire-coil (L1/L2) becoming permanently magnetized ... paralleling and performing the same electromagnetic characteristic of copper wire when it comes to magnetic field reformation (Rp ~ Rp1 ~ Rp2) of Figure (7-8), as further illustrated in electromagnetic coupling fields (71 ~ 511 ~ 512) of Figure (6-1) that encourages, brings-on, and perform Voltage Inducement Process (580) of Figure (6-1) as to (620) of Figure (7-1) without amp "influxing" (inhibiting amp flow) between Positive Voltage Potential (66) and Negative Voltage Potential (67) electrically applied across Resonant Cavities (140 ~170).

In-Line Circuit Components

Lengthening Inductor (L1/L2) lengths applies an even higher Voltage Potential (66/67) across Resonant Capacitor (140 ~170) (ER) since Inductance Reactance "Stores" Energy and, is expressed by

(Eq 19)

$$Wa = \frac{LI^2}{Z}$$

Where,
(Wa) is the energy in Joules (Watt-seconds); (L) is the Inductance in Henries; and (I) is the current in amperes.

Inductance Reactance directly determines "Stored" Energy (Wa) which is controlled by input Voltage Potential attenuated or varied by way of Voltage Amplitude (Vo xxx Va xxx V b ~ Vf xxx Vg xxx Vn) of Figure (7-13) and/or Gated Pulse-Frequency (49a xxx 49n ~ T3 ~ 49a xxx 49n), or both.

Stanley A. Meyer   7-6

RE: VIC Matrix Circuit   Memo WFC 426

      Inductance Reactance performs several functions simultaneously or to given stimuli: increases applied voltage amplitude (Vo ~ Vn), doubles input frequency (64a * 64b) when 50% Duty Cycle Pulse (T1 = T2) is inputted, effectuates "Step Charging Effect" (680) of Figure (7-7) when Pulse off-time (T2) is less than Pulse on-time (T1) ... determining voltage swing from highest voltage level (Vn) to volts switch-off point (Vff), and establishing Impedance (FL) which minimizes heat loss of electrical input power (49) by impairing electron movement.

      Inductor (L1) acts and performs in like manner to Inductor (L2) since both Inductor (L1/L2) are physically the same size and shape.

      Thermal Explosive Energy-Yield (gtnt)(16a xxx 16n) instantly produced from water (85) is determined by Voltage Amplitude (xxx Vn), Duty Cycle of Pulse Train (T1 ~ T2a xxx T1 ~ T2n), Gated Pulse-Frequency of applied Voltage Potential (49a xxxx 49n ~ T3 ~ 49a xxx 49n), Inductor (L1/L2) length, Secondary Pickup Coil (523) Length (FL3a xxx FL3n), dielectric gap-spacing (Cp), or any combination thereof.

Multi-layer Coil

      Inductance of a multilayer coil of rectangular cross section can be computed by below formula when optimizing maximum distributed capacitance (Cda xxx Cdn) and distributed inductance (DIa xxx DIn) of Figure (7-3) to intensify Inductance Field Strength (FLa xxx FIn) to function as a voltage multiplier in switch-off conditions (612a xxx 612n), as illustrated in (710) of Figure (7-10) as to VIC Coil Assembly (580) of Figure (6-1) and, is expressed:

(Eq 20)

$$L = \frac{0.8 \, (N \times A)^2}{6A + 9B + 10C} \quad \textit{Bobbin Cavity}$$

Where,
      (L) is the inductance in microhenries, (N) is the number of turns, (A) is the mean radius in inches, (B) is the length of the coil in inches, (C) is the depth of the coil in inches.

Stanley A. Meyer

RE: VIC Matrix Circuit   Memo WFC 426

Taper Resonant Capacitor (ERt)

Capacitor (ER) is automatically formed when dielectric liquid of water (Re) is placed between Electrical Conducting Plates (E1/E2) of Figure (1-1) page (1-13) (Memo WFC 420).

Stainless steel T304 material is used to form Electrical Voltage-Plates (E1/E2) which do "not" chemically interacts (chemically inert) (Lab tested less than .0001/year decomposition rate) with liberated water gases (hydrogen 86, oxygen 87, and non-combustible gases 74) being exposed to an high intensity voltage pulse-field (64a xxx 64n) with negligible amp flow. Electrical Plates herein called "Excitor" Plates or Voltage Zones (E1/E2) can take-on different configuration of shapes to maximize Dynamic Voltage Potential (600) of Figure (6-3) for different application of usage: (35a) Traveling Constant Electrical Voltage Wave by way of linear cylindrical resonant cavity (Tubular Cavity 730A), (35b) Traveling Compressional (concentrating electrical intensity) Electrical Voltage Wave by way of taper cylindrical resonant cavity (730B), (35c) Traveling Expanding Electrical Voltage wave by way of non-linear cylindrical resonant cavity (730C) ... or any voltage surface combination thereof ... each resonant cavity design acting and functioning as a Voltage Wave-guide (570) and gap-size (35) sufficient enough to allow the "Quenching Effect" to take place, as illustrated in (730) of Figure (7-12) as to (370) of Figure (3-40).

The dielectric property of water (being 78.54 ohms @ 25° C) permits the storage of "Electrical Charge" when a potential voltage difference exists between Electrical Voltage-Plates (E1/E2) as to (E9/E10).

Capacitance (Cp) of Figure (7-6) as to (690) of Figure (7-8) is determined by the surface area (A) of Electrical Voltage-Plates (E1/E2 ~ E9/E10), the distance (d) between the Electrical Plates (in inches), and the permittivity (Eo) of the dielectric property of water (85) and, is expressed in the following equation:

(Eq 21)

$$C = \frac{0.2249\, e\, A}{d\, Eo} \quad Picofarads$$

Stanley A. Meyer

RE: VIC Matrix Circuit   Memo WFC 426

Where,

(Eo) is Free-Space Permittivity of Water established by VIC Circuit (690) of Figure (7-8) ability to restrict amp flow, (e/Eo) Ratio is the Dielectric Constant of Water, (A) is the surface Area of Resonant Cavity and, is expressed in the below equation:

(Eq 22)

$$area\ (A) = \frac{h}{2}(a+b)\ Taper\ Resonant\ Cavity$$

Where,

(h) is longitudinal length of tapered resonant cavity, (a) being exit port circumference surface point (E9d) of Figure (6-2), (b) being cylindrical circumference surface point (E9a) of Figure (6-2) where tapered surface starts, (a)(b) circumference surface points (E9a)(E9d) respectively determined by below expressed equation: (see diagram 720 of Figure 7-11)

(Eq 23)

$$Circumference\ Surface\ Point\ (E9) = \pi D$$

Where,

(D) is diameter cross section of cylindrical surface at designated point (E9a ~ E9n), (π) being mathematical constant 3.1416.

Capacitance Reactance

Capacitance Reactance is determined by the insulation resistance (Rs+ Re) and Inductance (L1/L2) interacting together during D.C. Pulsing.

Dielectric property of water opposes amp leakage (Re) while another property of water takes-on an "Electrical Charge". Water temperature (Rt) (cool-to-the-touch) keeps (Re) constant since amp flow remains minimal. Plate Inductance (Lc) is Inductance Reactance of Inductor (L1)

Stanley A. Meyer   7 - 9

RE: VIC Matrix Circuit   Memo WFC 426

and Inductance Reactance of Inductor (L2) in series with Resonant Capacitor (140 ~170) of Figure (7-6) as to (690) of Figure (7-8).

In terms of Component Reactance, Inductors (L1/L2) should always be larger than Capacitor (ER) of Figure (7-2) in order to maximize amp restriction to enhance "Voltage Deflection" (SS' ~ 617a xxx 617n ~ RR') of Figure (7-4) and, is expressed by :

(Eq 24)

$$Z = X_L - X_C$$

Whereas,

Capacitor (ER) should remain relatively small due to the dielectric value of water to obtain maximum Thermal Explosive Energy-Yield (16a xxx 16n) of Figure (4-5) and subsequently establishing Quenching Circuit (370) of Figure (3-40) to prevent gas ignition inside traveling voltage wave-guide (590) of Figure (6-2) as to (730) of Figure (7-12) ... to bring-on and trigger Hydrogen Fracturing Process (390) of Figure (3-42) once liberated and expanding water gases (100) of Figure (4-8) passes beyond exit port (E9d) ... activating Voltage Ignition Process (90) of Figure (5-5) ... utilizing Dynamic Voltage Potential (600) of Figure (6-3) of opposite electrical stress (SS' ~ 617 ~ RR') to cause thermal atomic agitation (90) of Figure (4-7) (kinetic heat by atomic motion) which, when occurring at gas exit port (32) of Figure (4-5), spark-ignites expanding water gas-fuel (45/46/47) of Figure (4-5) during water inject cycle (70) of Figure (4-5) ... releasing thermal explosive energy (gtnt) (16) under control state.

Circuit Resistance

Total VIC "Circuit Resistance" to D.C. current flow is expressed and determined by:

(Eq 9)

$$Z = R_1 + Z_2 + Z_3 + R_E$$

Where,

(R1) is the resistive value of Secondary Pickup Coil (52) of Figure (7-8) plus Magnetic Field strength of primary coupling field (71) in direct relationship to inductance field strength (Rp) which is determined by the number of turns of wire that make up secondary coil-wrap (52), (Z2) is determined by inductance field strength (FL1) and resistive value (RS1) (typically 11.6 KΩ) of

Stanley A. Meyer   7 - 10

RE: VIC Matrix Circuit   Memo WFC 426

stainless steel (s/s) wire-coil (56) (L1) when being exposed to external magnetic coupling field strength (Rp), (Z3) is determined by inductance field strength (FL2) and resistive value (RS2) (typically 11.6KΩ) of stainless steel (s/s) wire-coil (62)(L2) when being exposed to the same external magnetic coupling field strength (Rp) ... each choke-coil (L1/L2) being of the same impedance value since both coil-wraps (56/62) are Bifilar wound together onto a single spool-bobbin, (Re) is the dielectric property of water and it's resistive value is typically (78.54 Ω) since "rain water" (85f) contains less than 20ppm of any type of contaminates due to Water Evaporation Process (530) of Figure (5-6). (see VIC Matrix Circuit 690 of Figure 7-8 as to Water Chart (760) of Figure (7-15), once again)

Transformer Action

Inductance Core (53) of Figure (6-1) composed of "Grain Oriented" Electrical Steel laminations step up applied Voltage (49) when Magnetic Field Coupling (71) of Figure (7-8) cross over to Secondary Pickup Coil-winding (52) which has more turns of wire than Primary Coil-winding (26) by way of "Eddy" currents that induces magnetic flux lines of forces (71a xxx 71n) emanating away from magnetic core material (53) and caused by Primary Coil (26) being electrically energized during pulsing operations (T1a xx T1n), as illustrated in (690) of Figure (7-8). Magnetic Induction (71a ~ 71n) is determined by Inductance Permeability (μL) of core material (53) along with VIC circuit geometry ability to step up Voltage Potential (Vo ~ Vn) by way of "Transformer Action", and is expressed in the following equations:

(Eq 25)

$$\frac{Ep}{Es} = \frac{Np}{Ns} \quad and \quad \frac{Ep}{Es} = \frac{Is}{Ip}$$

Where,
(Ep) is voltage induced in Primary Coil (26), (Es) is Voltage induced in Secondary Coil (52), (Np) is the number of turns of wire that make up Primary Coil-Wrap (504) of Figure (6-1), (Ns) is the number of turns of wire that make up Secondary Coil-Wrap (505) of Figure (6-1), (Is) is the established current flow (under load) in Secondary Coil-Winding (52), (Ip) is the amount of current flow in the Primary Coil-Winding (26) when electrically "energized" during pulsing operations (49a xxx 49n ~ T3 ~ 49a xxx 49a).

Stanley A. Meyer

RE: VIC Matrix Circuit   Memo WFC 426

The turns ratio of the VIC Transformer (26/ 52) is determined by the following equation:

(Eq 26)

$$T = \frac{Ns}{Np}$$

Where,

(Ns) is the number of turns of wire for each bobbin cavity (505) of Figure (6-1) as to (710) of Figure (7-10) that are electrically connected in series arrangement (505a xxx 505n) to form Secondary Coil-Wrap (52), (Np) is the number of turns of the primary Coil (26) wire-wrapped about spool cavity (504) ... each bobbin cavity adhering to equation (Eq 20), as illustrated in (710) of Figure (7-10).

The impedance ratio of VIC transformer is determined by:

(Eq 27)

$$Z = T^2$$

Where,

(T^2) is the sum of the magnetic field strength (FL4) of the primary coil (26) and the induced magnetic field (FL3) of the Secondary Pickup Coil (52) during each pulse cycle (T1) in direct relationship to repetitive pulse cycling (T1a xxx T1n) and both magnetic fields (FL3/FL4) interacting, and is expressed in the following equation:

(Eq 28)

$$M = \frac{La}{4}$$

Where,

(M) is the mutual inductance expressed in the same units as (La), (La) is the total inductance of Primary coil (26) and Secondary coil (52) with fields aiding.

Coupling Inductance (Rp) between the Primary coil (26) and Secondary Coil (52) is further extrapolated in the following equation:

(Eq 29)

$$Lt = \frac{1}{\frac{1}{L1+M} + \frac{1}{L2+M}} \quad Transformer$$

Where,

(Lt) is the total inductance, (L1 and L2) are the inductance of each individual transformer coils (26)(52), (M) is the mutual inductance of each transformer coil (26/52) being in parallel relationship with fields aiding.

Coupling Inductance (Rp1) and (Rp2) in (690) of Figure (7-8) is further expressed in the following equation:

(Eq 30)

$$Lt_{CC} = L1 + L2 + 2M \quad Choke\ Coils$$

Where,

(Lt_{CC}) is the total inductance of Choke Coils (FL1 ~ FL2), (L1) and (L2) are the inductance of each individual choke coil (56)(62) in series with Secondary Coil (52) Electrical Voltage Potential (700) of Figure (7-9) and being exposed to the same Voltage Transformer (26 ~ 53 ~ 52) magnetic field (Rp) with aiding fields, (M) is the mutual inductance of choke coils (L1/L2) since Transformer Magnetic Field (Rp) is the excitation External Magnetic Field (Rp1/Rp2) by way of Unipolar Pulsing Core (53).

VIC Coil Assembly (580) of Figure (6-1) as to (690) of Figure (7-8) in reference to Schematic Circuit (620) of Figure (7-1) is constructed in such a way as to rotate and position Inductor Coils (26 ~ 52 ~ 56 ~ 62) to be of the same electromagnetic polarity orientation, indicator mark (•) ... thus, allowing Inductance Fields (FL1 ~ FL2 ~ FL3 ~ FL4) to be aiding one another during the same sequence of pulse-time (T1) ... thereby, allowing Inductance Charging Effect (660) of Figure (7-5) and Resonant Voltage Effect (670) of Figure (7-6) to interact with the dielectric properties of water (Re) to cause and inhibit electron flow (I_F) since "electrons" magnetic

RE: VIC Matrix Circuit   Memo WFC 426

field (547) of Figure (5-9) locks onto the electromagnetic fields of each energized choke coils (FL1/FL2) during Voltage Excitation (Vo ~Vn) which, now, brings on and allows "Electron Bounce Phenomenon" (700) of Figure (7-9) to take place. *(See Appendix B Note 1)*

Electron Bounce Phenomenon

High Voltage Potential of Difference (Vo ~ Vn) (SS' ~ 617 ~RR') is accomplished when magnetic flux lines of force (71a xx 71) (Rp) emanating away from closed-loop magnetic pulsing core (53) of Figure (190) penetrates Inductance coil-windings (52 ~ 56 ~ 62) simultaneously during each and every pulse on-time (T1a xxx T1n) as programmable pulse-train (49a xxx 49n ~ T3 ~ 49a xxx 49n) is adjusted to "Tune - in" to the dielectric property of Water (Re) .... causing mutual inductance (μl) (see equations Eq 28 thru Eq 30) to transform Distributed Capacitance (Cda xxx Cdn) of Figure (7-3) of each inductance coils (52 ~ 56 ~ 62) into a coherent Voltage Potential (Vo ~ Vn) equaling the sum of Voltage Potential (Vp) developed across each Pickup Coils (VpT+ Vp1 + Vp2) ... producing Dynamic Voltage Potential (600) of Figure (6-3) during repetitive pulsing (49a xxx 49n ~ T3 ~ 49a xxx 49n) ... setting up and performing pulsating Opposite Electrical Attraction Fore (SS' ~ 617 ~ RR' - T3 - SS' ~ 617 ~ RR') of Figure (7-4) as to Voltage Dynamics (220) of Figure (3-29) ... triggering Hydrogen Fracturing Process (90) of Figure (5-5) as to (100) of Figure (4-8) ... instantly releasing thermal explosive energy (gtnt) (16) from Water (85) on demand, as illustrated in Taper Resonant Cavity (590) of Figure (6-2) as to (70) of Figure (4-5). The resultant Dynamic Voltage Potential of Difference (opposite electrical attraction force) (SS' ~ 617 ~ RR') is in balance phase of equal electrical intensity (66 = 67) of opposite polarity (positive electrical voltage potential 66 equals negative electrical voltage potential 67) since the Voltage Coefficient of Inductance (FL1/FL2), Voltage Coefficient of Capacitance (Cd1/Cd2), and Voltage Coefficient of Resistance (Rs1 / Rs2) across choke coils (L1/L2) are the same values ... allowing, Voltage Bounce Phenomenon (700) of Figure (7-9) to be preformed.

Magnetic Field Coupling (71) of Figure (7-9) entering into and passing through Secondary Coil-winding (52) of Figure (7-8) causes and produces copper ions (643a xxx 643n) (Positive Charged atoms 542a xxx 542n having missing electrons) when moving external electromagnetic field strength (71a xxx 71n) is sufficient enough to dislodge electromagnetically charged electrons (641a xxx 641n) from copper atoms making up copper wire material (52). Collectively, the resultant positive electrical charged copper ions (642a xxx 642n) added together produces Positive Voltage Potential (629) being electrically applied to choke-coil (56); whereas, the "Liberated" negative electrical charged electrons (641a xxx 641n) added together provides Negative Voltage

RE: VIC Matrix Circuit   Memo WFC 426

Potential (631) to the opposite end of Secondary Wire (52) being electrically connected to choke coil (62). Once Secondary Coil-winding (52) is de-energized by the removal (collapsing magnetic field during pulse off-time T2) of external Magnetic Field (71), the dislodged electrons (641a xx 641n) return to positive charged copper ions (642a xx 642n) ... terminating and switching off opposite voltage potential (629 ~ 631) when positive electrical state of the copper atoms changes back to net electrical charge of zero. Sustaining and maintaining the resultant induced Voltage Potential (Vo ~ Vn) without "Electron Discharged" (inhibiting electron flow) through Choke Coil (62) while, at the same time, inhibiting (preventing) any additional or other electrons from entering into Secondary copper wire-zone (52) by way of Choke Coil (56) is herein called "Electron Bounce Phenomenon" (EbP), as illustrated in (700) of Figure (7-9).

Electrically Interlinked serially together, Electron Bounce Phenomenon (EbP), Voltage Coefficient of Inductance (Fl1/Fl2), Voltage Coefficient of Capacitance (Cd1/Cd2), Voltage Coefficient of Resistance (Rs1/Rs2), and dielectric Coefficient of Water resistance (Re) allows Voltage Potential (Vo ~ Vn) of opposite electrical polarity to perform work (SS' ~ 617 ~ RR') without amp influxing ... thus, not allowing the introduction of electron flow into Hydrogen Fracturing Process (90) of Figure (5-5) during Voltage Stimulation (SS' ~ 617 ~ RR') ... causing "electron clustering" (641a xxx 641n) to take place within Copper Wire Zone (52) during pulse on-time (T1) ... inhibiting "electron flow" to maintain opposite voltage potential (66/E9 ~ 67/E10) across Resonant Water Gap (616) during the process of converting water-fuel (85) into instant thermal explosive energy (gtnt) ... therefore, producing a physical force-yield (Fy) during gas-ignition (70) of Figure (4-5) which is directly related to the liquid volume of water (85) per injection cycle and applied Resonant Voltage Intensity (Vo ~Vn), as illustrated in (590) of Figure (6-2) as to (90) of Figure (5-5).

Of course, in practical terms of component interaction, a minute amount of amp leakage is present and does occur due to Electronic Component Limitations but is negligible as to the overall performance of the Hydrogen Fracturing Process (590) of Figure (6-2) when being subjected to either one of traveling Electrical Voltage Wave-forms (730a ~ b ~ c) of Figure (7-12), see Voltage Graph (750) of Figure (7-14) once again.

Voltage Amplitude Switch-Off

Voltage levels of variance (Va xxx Vn) is achieved by simply switching-in or switching-out the member of Secondary Coil-cavities (505a xx 505n) (see 740 of Figure 7-13) in direct

Stanley A. Meyer

RE: VIC Matrix Circuit   Memo WFC 426

relationship to Taper Resonant Voltage surfaces (E9/10) of Figure (6-2) which acts and performs as a "Voltage Amplifier" when Compressional Wave-form (B) of Figure (7-12) is intensified at Exit Port (32) of Figure (6-2). Switching the member of Secondary Coil-Array (505a xxx 505n) maximizes electrical power transfer from Primary Coil (26) to Secondary Coil (52) by keeping Voltage Amplitude of Pulse-train (49a xx 49n ~ T3 ~ 49a xxx 49n) constant.

Mode of Operability

The established "mode-of-operability" of VIC Coil Assembly (580) of Figure (6-1), now, allows Voltage Potential (Vn) of opposite voltage polarity (66/SS' ~ 67/RR') to increase and be attenuated up to and beyond 20 Kilovolts while inhibiting and restricting amp leakage in the milliamperes range ....establishing operational parameter of utilizing Opposite Electrical Attraction Force (SS' ~ RR') of high voltage intensity (Vn) to instantly release thermal explosive energy (gtnt) from natural water. Voltage Compressional Wave-form (35b) and Expanding Voltage Wave-form (35c) increases the intensity of applied pulsating opposite electrical attraction force (SS' ~RR'a xxx SS' ~ RR'n) even further during each new pulse-cycle (T2 next T2) across water-gap (616) ... increasing Thermal Explosive Energy-yield (gtnt) to higher energy-levels (gtnta xxx gtntn) beyond applied excitation voltage (Vn) by simply altering Voltage Surfaces (35b/35c) as in reference to Linear Voltage Surfaces (35a), as illustrated in (730) of Figure (7-12). Pulse Off-time (T2) of Figure (7-8) as to (620) of Figure (7-1) is adjusted to compensate for the rise and fall of magnetic coupling field (71) to produce applied Unipolar Wave-forms (64a xxx 64n) entering into Wave-guides (35a/35b/35c). Less water contaminants nets even higher energy-yield (gtnta xxx 85a ~ 85h xxx gtntn), as illustrated in Water Chart (760) of Figure (7-15). In terms of thermal explosive energy-yield (gtnt) under dynamic pressure of compression approximately 7.4 (μl) microliter of a liquid-volume of a water droplet per injection cycle is all that is required to run the Dune Buggy 1600cc 50hp VW I.C. engine at 65 m.p.h. on the open road; whereas, a typical 325 hp diesel I.C. truck-engine would require about 48.1 (μl) microliters of a water droplet per injection cycle to accomplish the same open road performance. (see WFC Water vs Gasoline Energy Content Equations (memo WFC 429).∆

Stanley A. Meyer   7 - 16

RE: VIC Matrix Circuit   Memo WFC 426

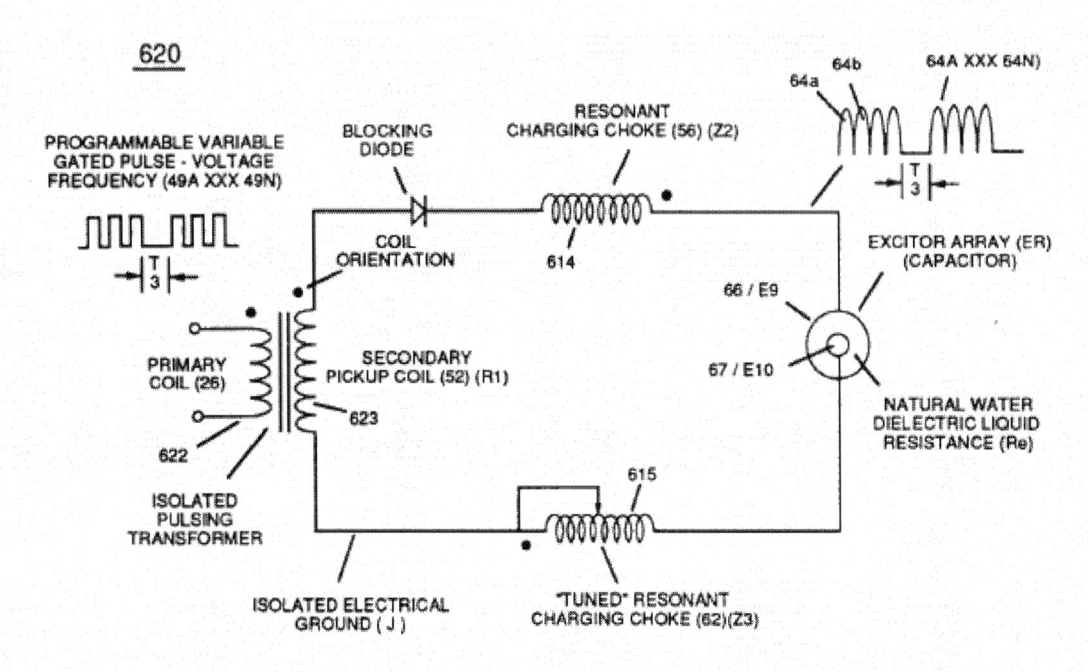

FIGURE 7-1: VIC IMPEDANCE NETWORK

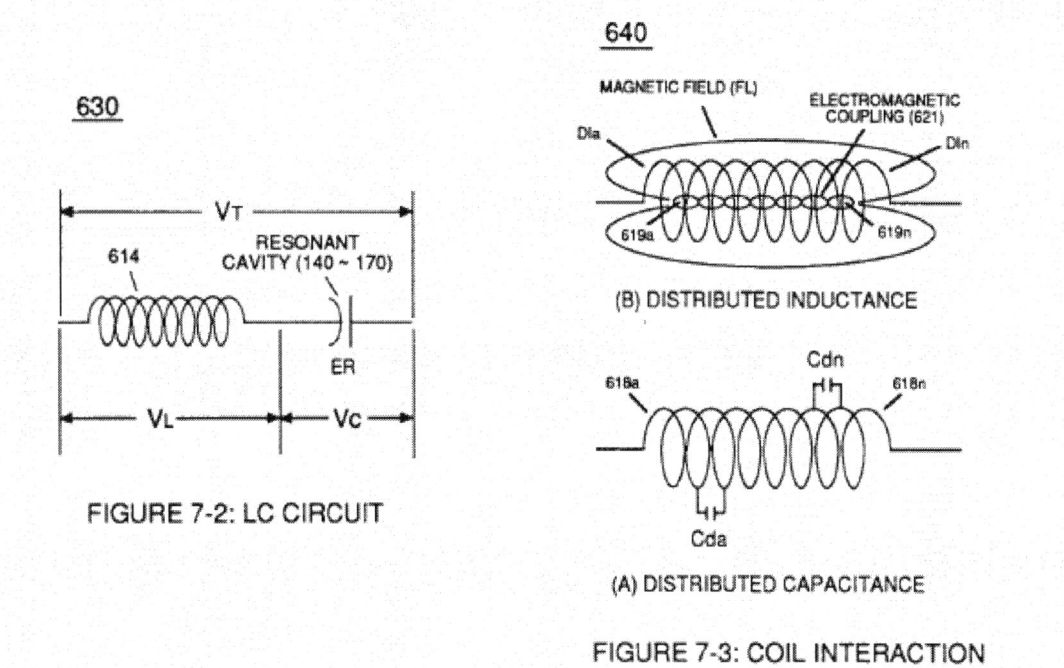

FIGURE 7-2: LC CIRCUIT

(B) DISTRIBUTED INDUCTANCE

(A) DISTRIBUTED CAPACITANCE

FIGURE 7-3: COIL INTERACTION

Stanley A. Meyer   7-17

RE: VIC Matrix Circuit   Memo WFC 426

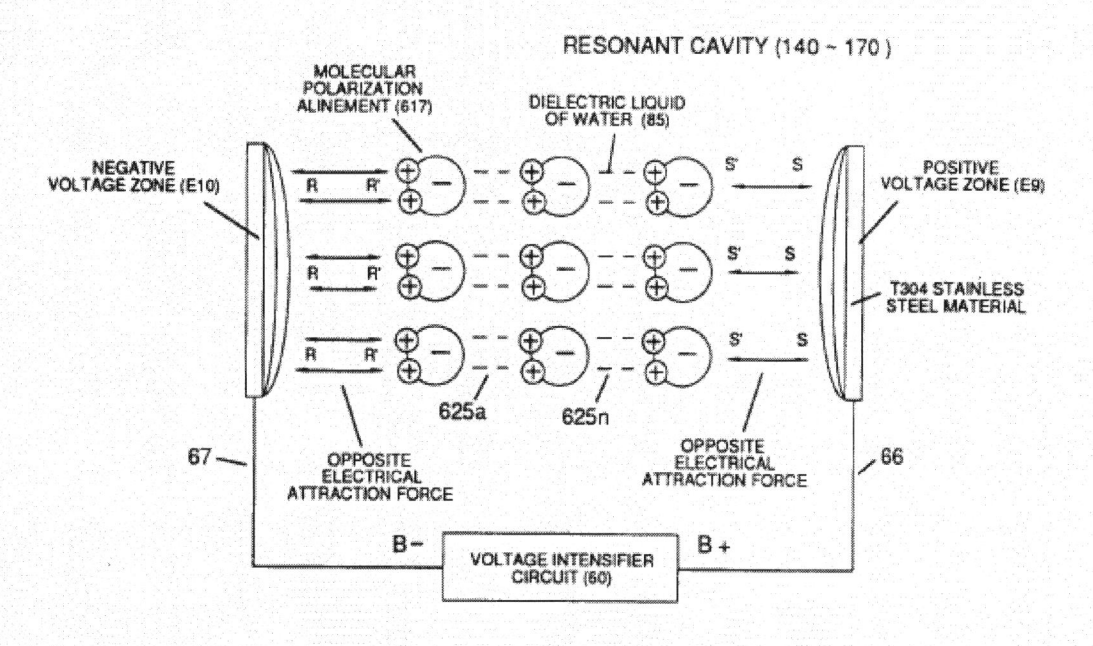

650

FIGURE 7-4: ELECTRICAL CHARGING EFFECT

660

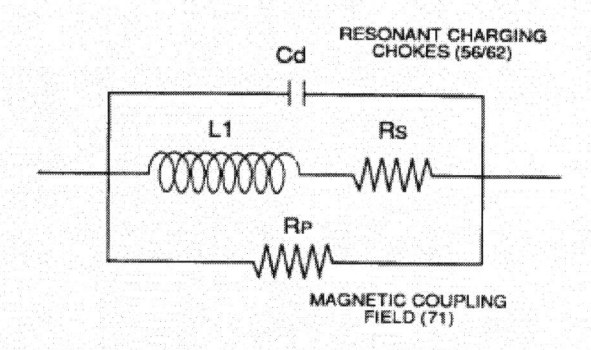

FIGURE 7-5: INDUCTANCE CHARGING EFFECT

Stanley A. Meyer   7-18

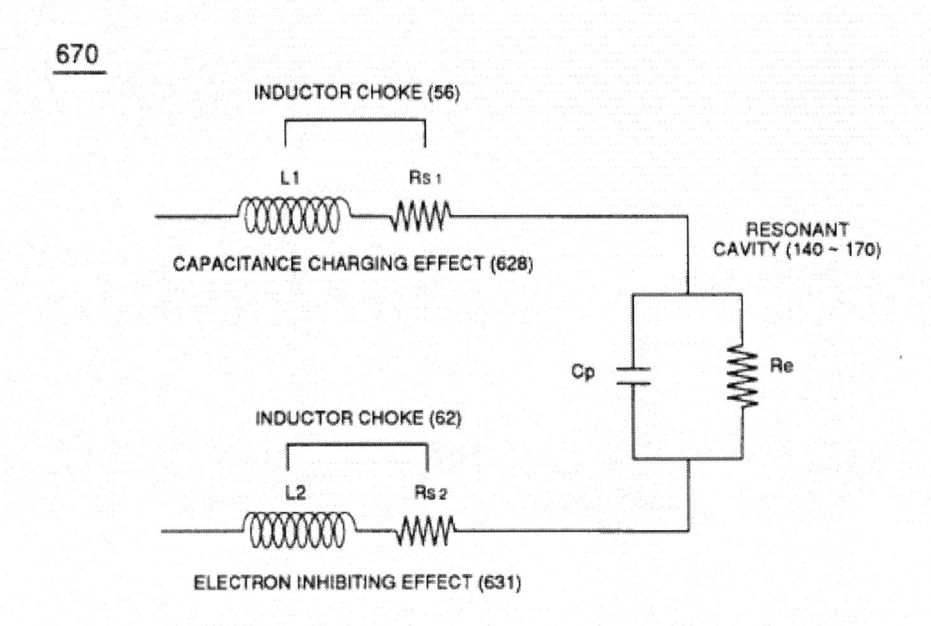

FIGURE 7-6: RESONANT VOLTAGE EFFECT

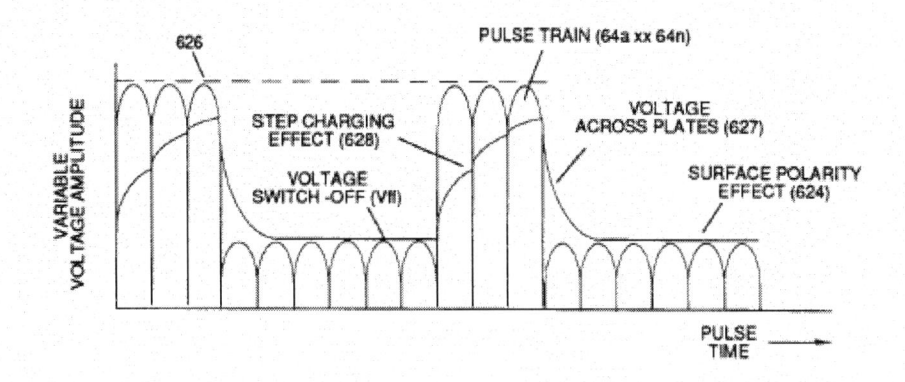

FIGURE 7-7: VOLTAGE CHARGING EFFECT

RE: VIC Matrix Circuit   Memo WFC 426

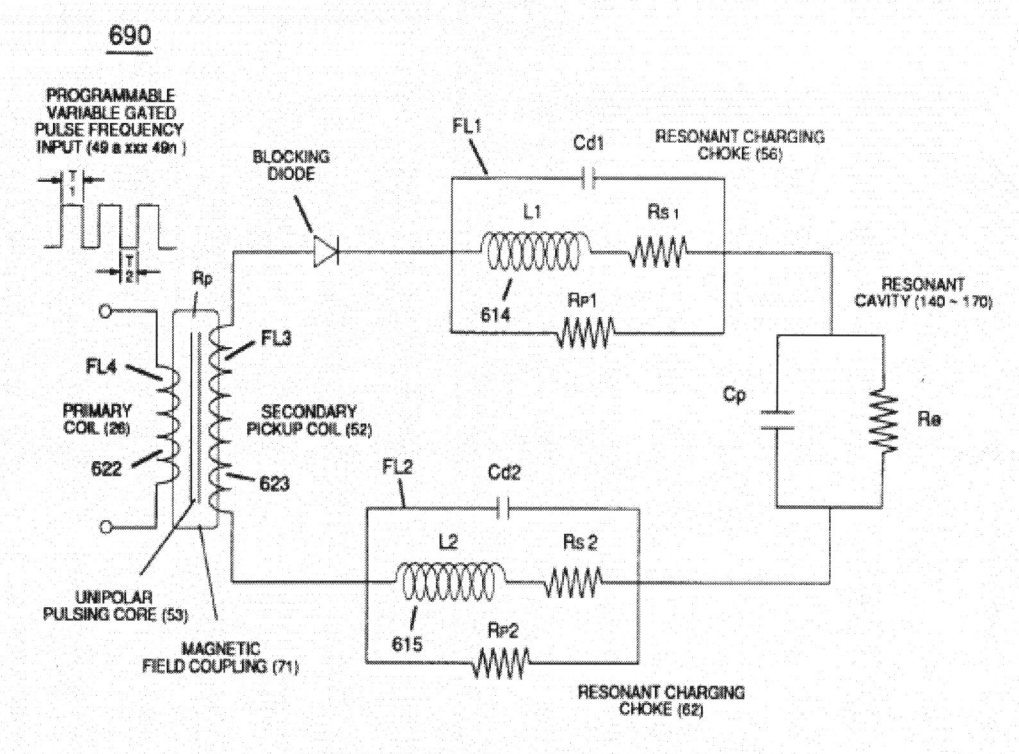

FIGURE 7-8: VIC MATRIX CIRCUIT

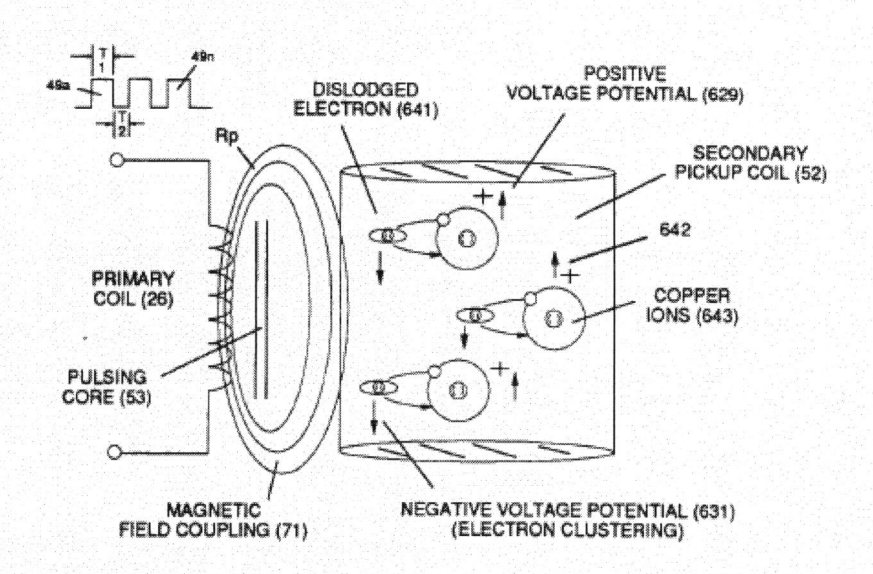

FIGUREM 7-9: ELECTRON BOUNCE PHENOMENON (EbP)

Stanley A. Meyer   7-20

RE: VIC Matrix Circuit   Memo WFC 426

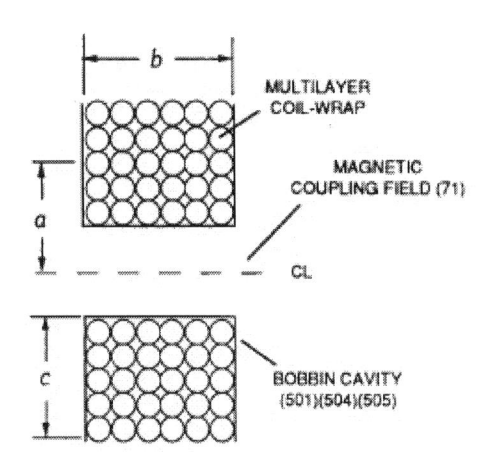

710

FIGURE 7-10: VOLTAGE STEPPING COILS

720

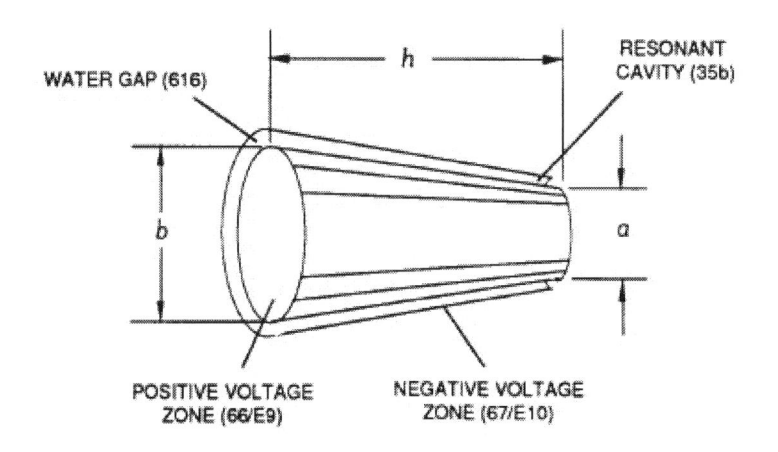

FIGURE 7-11: TAPERED VOLTAGE WAVE-GUIDE

RE: VIC Matrix Circuit

Memo WFC 426

730

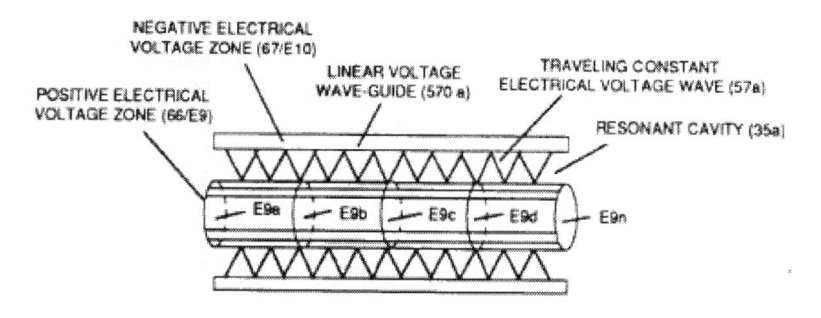

(A) LINEAR CYLINDRICAL RESONANT CAVITY

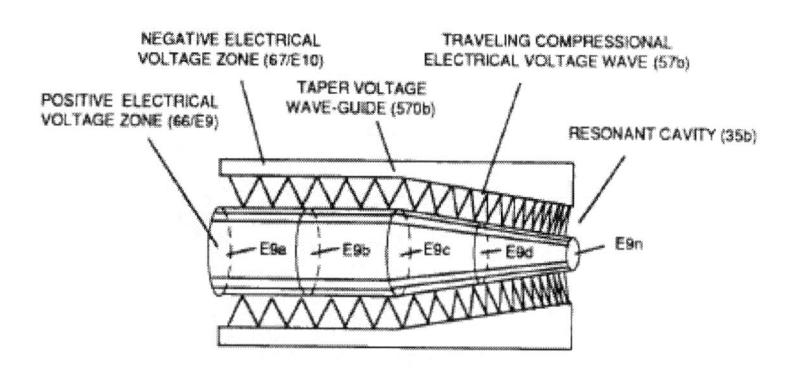

(B) TAPER CYLINDRICAL RESONANT CAVITY

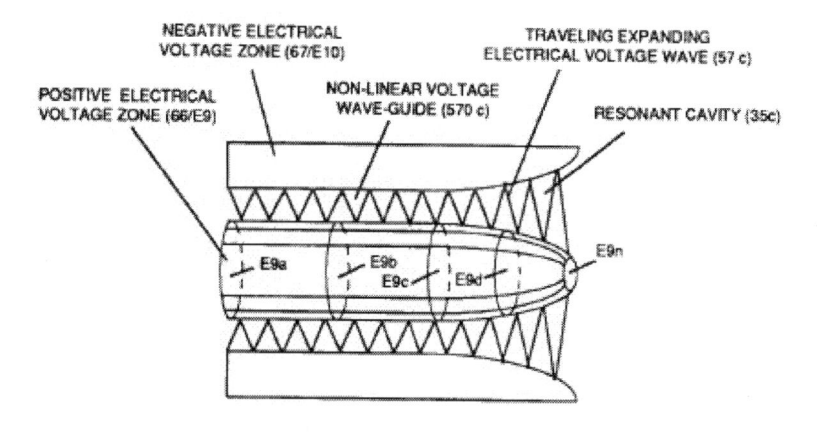

(C) NON-LINEAR CYLINDRICAL RESONANT CAVITY

FIGURE 7-12: RESONANT CAVITY ELECTRICAL VOLTAGE WAVE

Stanley A. Meyer

7-22

RE: VIC Matrix Circuit　　　　　　　　　　　　　　　　　　　　　Memo WFC 426

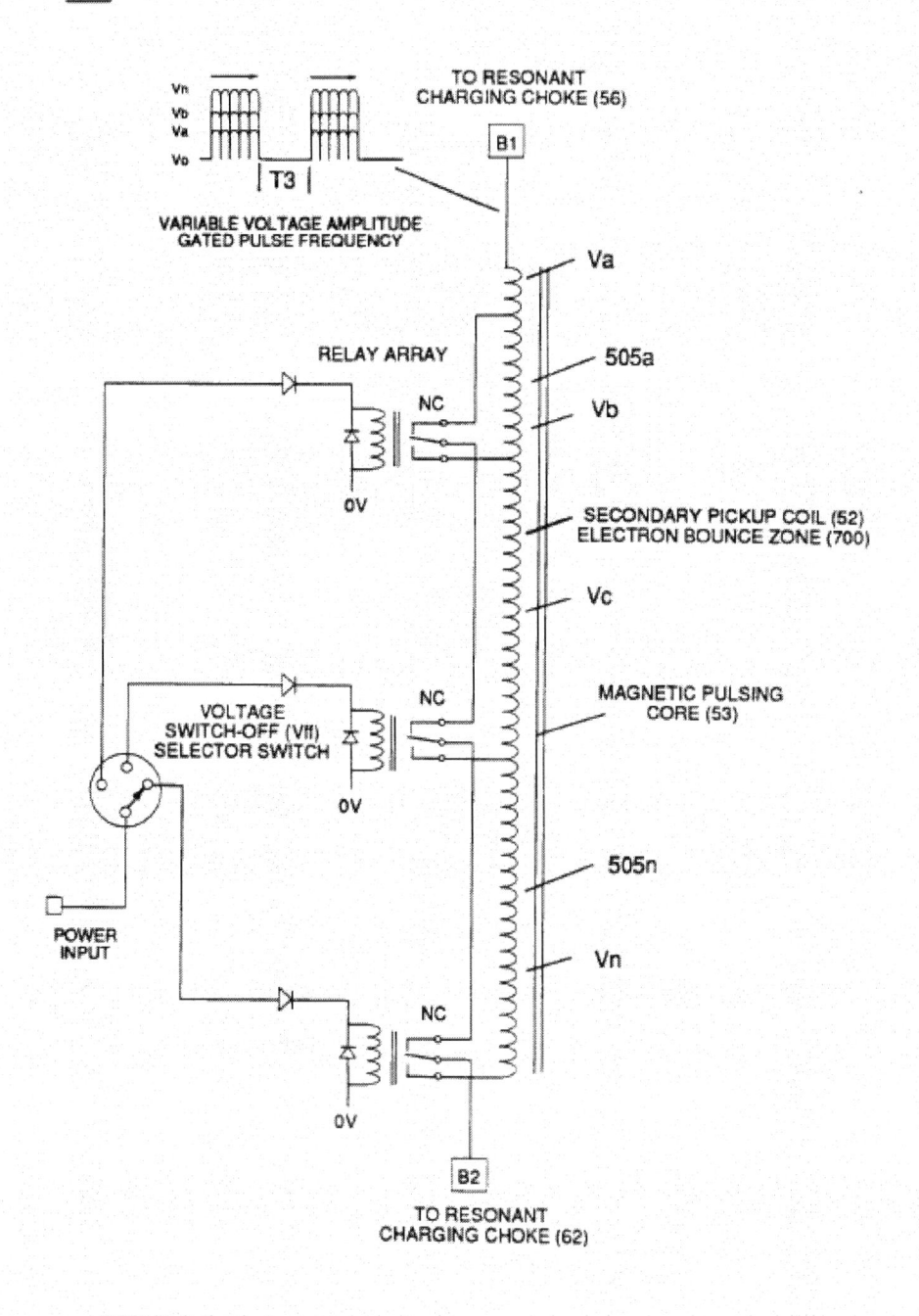

FIGURE 7-13: VIC SECONDARY SWITCH-OFF COIL-ARRAY

RE: VIC Matrix Circuit　　　　　　　　　　　　　　　　　　　　　Memo WFC 426

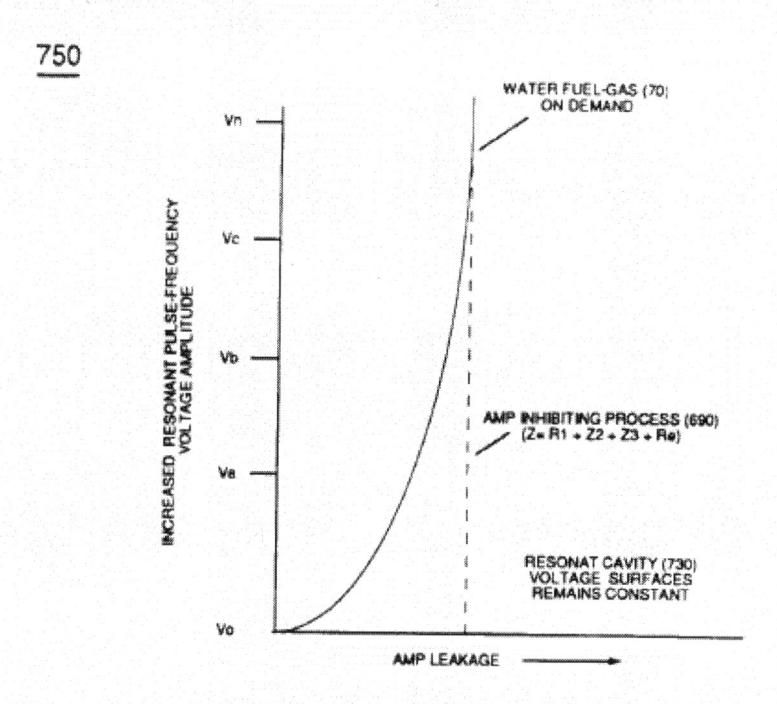

FIGURE 7-14: RESONANT CAVITY WATER-FUEL INJECTION

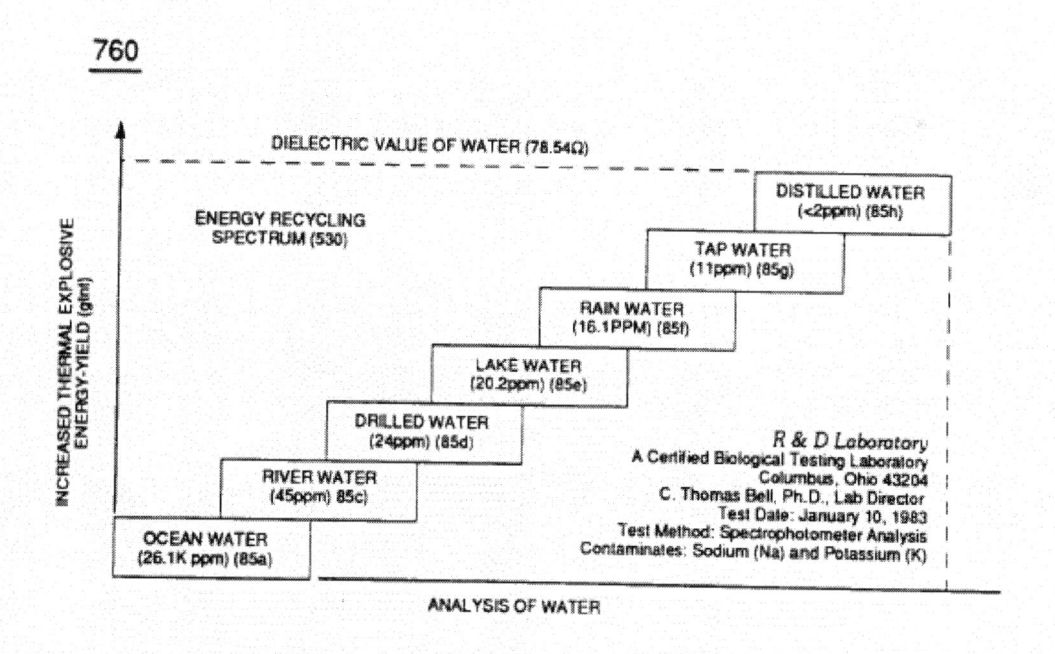

FIGURE 7-15: THERMAL EXPLOSIVE-ENERGY OF WATER

Memo WFC 427

Voltage Wave-Guide

Propagating "Resonant Action"

By Voltage Tickling of State Space

The "Mode of Operability" of determining the "Operational Parameters of adjusting thermal explosive energy (gtnt) exiting Water Fuel Injector nozzle-port is directly related to the type of Voltage Pulse Train being used and the geometrical configuration of the Resonant Cavity.

In terms of Voltage Pulse Wave-form, several Electrical Operational Parameters exists: Dynamic State Space which varies Electrical Stress Intensity continually during Unipolar Voltage Pulse formation and Static State Space being an electrical condition by which Electrical Stress is being held constant during a certain time-period.

In reference to Resonant Cavity design configuration to bring-on Energy Vectoring of the Hydrogen Gas Flame Front, the following Water Fuel Injectors are utilized: Linear Cylindrical Resonant Cavity; Taper Cylindrical Resonant Cavity; and Non-Linear Cylindrical Resonant Cavity.

Combining "Electrical Voltage Parameters" with "Physical Design Parameters" of a given type of Resonant Cavity allows the energy-yield of the Hydrogen Gas Flame Front to be either more or less Thermal Explosive Energy (gtnt) over thermal heat energy and herein is called "Energy Vectoring" ... and is performed in the following way:

Section 8

RE: Voltage Wave-Guide   Memo WFC 427

Voltage Wave-Guides

Propagating

"Resonant Action" By Voltage Tickling of State Space

The "mode-of-operability" of determining the "operational parameters" of releasing thermal explosive energy (gtnt) (flame force-yield) in direct relationship to thermal heat-yield (flame temperature) beyond applied voltage pulse-frequency of opposite polarity intensity (farthest from state of equilibrium) is effectuated by simply changing or altering the physical configuration of the "Voltage Wave-guides" (Resonant Cavity structure of design) in direct reference to applied ever changing Unipolar Voltage Amplitude Pulse-Wave (see Figure 6-3 of WFC memo 425) to bring-on and propagate a predetermine flame-density of projection by preselecting the "State Space" in which the injected and incoming water bath is subjected to ... performing work of instantly converting water into thermal explosive energy (gtnt) on demand...using water as fuel. The established "State Space" is governed by either one of two variables being of either "Static" or "Dynamic" state of condition. Dynamic State condition is a variable of condition that is changing all the time; whereas, Static variable of condition is set at some point but then never changes. A point in "State Space" represents the state of the flame-system at a given time and is either in an synchronized and synchrony movement. Interaction of either " Static State Space" or "Dynamic State Space" in similar state of conditions or opposite conditions of "states" allows "Energy-Vectoring" of the hydrogen flame-front or flame projection of trajectory ... establishing an interrelationship on how to design retrofit the water fuel injector ® to any existing type of energy consuming device.

Static variables condition is established when the resultant gas pressure is held constant with the never changing static electrical stress of opposite polarity of Voltage Pulse-Wave (64) of Figure (3-20) of WFC memo 422DA. Dynamic variable conditions exists when both the applied electrical stress of opposite polarity and dynamic gas pressure are continually changing in a preset time frame. Where, Combinatorial variables conditions is a "State Space" function of subjecting constant static gas pressure to an ever changing electrical stress of opposite polarity (RR'-SS'). And, Differential dynamic variables condition of "State Space" is accomplished whenever changing dynamic electrical stress of opposite polarity (RR'-SS') encounters a negative (decrease/drop-in) dynamic gas pressure. In each and all "Space State" of changes, the combustible gas atoms of water is/are "Electrically Stress" under "different" pressure levels to bring-on the triggering point of thermally igniting the combustible gases of water beyond or away from "Stable State" of Equilibrium. Voltage Tickling of State Space under "Resonant Electrical Stress" without amp influxing while "Tuning-In" to the dielectric properties of water is herein referred to in this WFC Tech-manual as "Resonant Action."

Traveling Voltage Wave-Guides

The formation of tubular Traveling Voltage Wave-guide (570a) of Figure (7-12) (WFC memo 426) as to (770) of Figure (8-1) is physically formed when positive electrical voltage surface (66/E9) and negative electrical voltage surface (67/E10) are placed in parallel space relationship to form voltage surfaces (E9/E10) about an cylindrical axis of rotation having space-gap (35) there

RE: Voltage Wave-Guide   Memo WFC 427

between... and thus, forming Cylindrical Resonant Cavity (730A) of Figure (7-12) as to (770A) of Figure (8-1) when space-gap (616) of Figure (720) exposes injected water bath (85) to unipolar pulse-oscillation of high voltage intensity of opposite polarity (67/66) as to (780) of Figure (8-2) which, in turn, propagates opposite electrical attraction force (RR' ~ SS') of Figure (7-4), as illustrated in (590) of Figure (6-2) as to (585) of Figure (8-1). The dielectric property of water (85) (resistance to electron flow) in conjunction with VIC Coil Matrix Circuit (690) of Figure (7-8) (WFC memo 426) as to VIC Coil Assembly (580) of Figure (6-1) (WFC memo 425) ability to inhibit amp "influxing" (Electron Bounce Phenomenon EbP) during pulsing operations (49a xx 49n) allows voltage amplitude of pulse-frequency potential (T1a xxx T1n) as to (Vo ~64a ~64b ~64c ~ Vn) of (780A) Figure (8-2) to be applied across cross-sectional circular-ring water bath (85) (donut shape) to cause Voltage Wave-Form (57) of Figure (6-2) to travel the entire longitudinal length of water-gap (616) since stainless steel material (s/s) (T304) forming Voltage surfaces (E9/E10) electrically conducts and transfers (skin effect) Voltage Pulse-Frequency Potential (583) alone the inside surface area of the chemically inert and non-oxidizing stainless steel (s/s) tubular material (E9/E10) which physically dictates the shape and configuration of voltage zones (66/67)... forming tubular voltage wave-guide (s) (570) of Figure (7-12) that, now, becomes the same physical configuration of Water Gap (616), as illustrated in (720) of Figure (7-11).

    The surface tension of water (584) adjacent to both voltage surfaces (E9/E10) further aids the transmission of voltage potential (66/67) since Electrical Charging Effect (585) of Figure (7-4) does not change or alter the dielectric value of water (Re). Together, the Voltage Coefficient of Water (e/Eo) of Equation (Eq 21) and the Voltage Coefficient of the stainless steel (s/s) material forming voltage surfaces (E9/E10), now, allows the establishment and performance of Traveling Electrical Voltage Wave-Guide (583/602) since electrical conductance zone (587) between electrical surface (S) (E9/E10) and the dielectric surface tension of water (584) acts and performs as a electrical conductor (Skin Effect) ... since electrical transmission zone (587) is almost free of electron leakage ... since Water Bath (85) is a dielectric-liquid (typically 78.54Ω) that does not like to transfer nor exchange electrons ... thereby, maintaining voltage amplitude potential (V0 ~ 64a ~ 64b ~64c ~ Vn) of Figure (8-6) without experiencing amp arc-over across Water-Gap (616) in any appreciable amount ... allowing pulsating opposite electrical attraction forces (RR' ~ SS') to perform the work of "Electrically Charging" water bath (85) to bring-on and trigger Hydrogen Fracturing Process (90) of Figure (5-5), as illustrated in Energy Pumping Stage (520) of Figure (5-3). Voltage Intensifier Matrix Circuit (690) of Figure (7-8) electrically connected with resistive-liquid (85/Re) (forming Resonant Water Gap C_p of Figure 7-8) propagates the transmission of Traveling Voltage Wave-Form (57) of Figure (6-2) as to (770) of Figure (8-1) by the functional

RE: Voltage Wave-Guide   Memo WFC 427

relationship of Circuit Resistance Equation (Eq 9) during programmable Voltage Pulsing operations (49a xxx T3 xxx49n) of Figure (8-2).

Electrical Voltage-Pulse Wave-Transmission (583a xxx 583n), now formed, occurs along Electrical Conductance Zone (587) since applied Electrical Pulse Voltage amplitude (Vo ~ 64a ~ 64b ~ 64c ~ Vn) is time responsive (T1/T2a ~ T3 ~ T1/T2n) to incoming gated Voltage Pulse Frequency (49a xxx ~ T3 ~ xxx 49n). Each Voltage Pulse duration time-period (T1 on time) from start to finish is directly related to applied Voltage-Pulse Amplitude (Vo xxx Vn) and reoccurring Voltage Pulse Frequency (49a xxx 49n) forming "Unipolar Voltage Pulse-Wave" (583) from zero voltage ground state (Vo) to a predetermined Voltage Level (xxx 64 x ~ 64y ~ 64z ~ Vn) on the leading edge of the Voltage Pulse-Wave (Vpa) and, then, reversing voltage up swing to drop on the trailing edge (Vpb), completing Voltage-wave (583). The newly established leading voltage edge (Vpa) and trailing voltage edge (Vpb) being uniform in shape/configuration since both Resonant Charging Chokes (56/Z2 ~ 62/Z3) resistive values are the same (Typically 11.6KΩ each) and incoming signal (49a xxx 49n) is electrically linked with Water-Gap Capacitor (Cp) of Figure (7-8) having dielectric liquid of Water (85) there between ... Thereby, preventing coil-ringing during each pulse off-time ... allowing Electron Bounce Phenomenon (EbP) to occur without amp influxing within VIC Matrix Circuit (690) of Figure (7-8) as so governed by Circuit Resistance Equations (Eq. 9) which, in activated electrical-state, allows positive Voltage Pulse-Wave (583) to be duplicated in succession to form Voltage Pulse Train (66 ~ 583a xxx 583n), as illustrated in (770) of Figure (8-1). Opposite negative Voltage Pulse Train (67 ~ 602a xxx 602n) is similarly formed since "Electron Clustering Effect" (631) of Figure (7-9) produces a "Negative Electrical Voltage Intensity (67) in equal magnitude to the "Positive Electrical Voltage Intensity (66) during each/repetitious magnetic pulse-cycle (Rp/71). Remember, Secondary Voltage pickup coil (52) of Figure (7-8) displaces and separates Resonant Charging Chokes (56/62) on opposite end of said Secondary Pickup Coil (52).

State Space (Sp)

During the electrical-formation (66-Vpa/Vpb ~ 67- Vpa/Vpb) of each opposite Electrical Voltage Pulse-Wave (66-583 ~ 67-602), opposite electrical attraction force (RR' ~ SS') of Figure (7-4) is produced across water cap (Cp) of Figure (7-8) which, now, sets up and defines the conditions of "State Space," as illustrated in (770A/B) of Figure (8-1) as to (650) of Figure (7-4). The newly formed Opposite Electrical Attraction Force (RR' ~ SS') intensity is directly related to the applied Voltage Amplitude Burst-Time (Vpa ~ Vn ~Vpb) as to the Voltage Burst-Frequency (49a xxx 49n)

Stanley A. Meyer

RE: Voltage Wave-Guide   Memo WFC 427

as to Voltage Peak Excursion Point "P" at the height of Unipolar Voltage Pulse Wave (583/602) which, in turns, determines maximum Voltage Peak-Potential (Vpp) at any given time during each Voltage Pulsing Cycle (Vpa/Vpb). Electrical Attraction Force intensity (RR' ~ SS' as to RU/RU' ~ ST/ST' as to 550 of Figure 5-8) at Peak Voltage Potential (Vpp) is either increasing or decreasing or remaining constant as to Voltage Peak Excursion Point "P" trace-position which scans the exact Voltage Pulse Wave-Form (66-583/67-602) being produced, as illustrated in (780B) of Figure (8-2). Equalizing Voltage Pulse-Scan (Eps) from start of one Voltage Pulse-Field (Vpf) to the start of next Voltage Pulse-Field (Vpfa + Vpfb + Vpfc + Vpfn) is determined by the total average of the number of applied Voltage-Pulses (Vp ave.) making up opposite Voltage Pulse Train (583/602a xxx 583/602n) in synchronous movement. Generally speaking, Arc Curve (Vac) changing/varying to Arc-Line (Val) of Unipolar Voltage Pulse Wave-form (Vpwf) defines Voltage Pulse Field (Vpf) scan profile (Vsp) by which Trace-point "P" determines the type of "State Space" being used to propagate "Voltage Tickling" of water molecule (85) undergoing "Electrical Stress" under different fluid-pressures.

Whenever, Voltage Excursion Point "P" is always changing in a given space-time, "State Space" is referred to as "Dynamic State Space;" whereas, "Static State Space" exists when Voltage Excursion Point "P" remains constant during a precise period of space-time at Peak Voltage Potential (Vpp) forming clipped Voltage Pulse Wave-form (Vcwf) during Voltage Pulse Shaping by way of Programmable Pulsing Circuit (WFC project 422DA/423DA) electrically interfaced with VIC Matrix Circuit (690), as illustrated in (780) of Figure (8-2). Dynamic State Space causes Opposite Electrical Attraction Force (RR' ~ SS') to continually vary in electrical intensity (RR' ~ SS' vei) as to formation of Voltage Peak Curve (Vpc); wherein, Static State Space allows Opposite Attraction Force (RR' ~SS') to remain at constant electrical intensity (RR' ~ SS'cei) when Peak Voltage Potential (Vpp) is clipped in forming Arc-Line (Val), as illustrated in (780C) of Figure (8-2). Crossover Unipolar Pulse Train (780B) is used when particle oscillation of the water molecule atom (s) is/are to be continually electrical stressed (RR' ~SS' vei) under changing conditions of higher magnitude (Compressing Voltage Pulse Wave-form) than the use of Planar Unipolar Pulse Train (780A). Clipped Unipolar Pulse Train (780C) is used to encourage further increase in atomic dwell-time capable of raising Atomic Energy Level (AEl) of the Water Atoms to even a higher energy-state before Snapping-Action occurs when Unipolar Pulse Wave (Upw) returns to ground-state (Vo) after voltage propagation (Vpa/Vpb). Of course, the repetition-rate of "Atomic Snapping-Action" (Asa) (the number of Voltage Pulse Fields Vpf occurring per unit of space-time) directly determines the resultant energy level of Static Electrical Charging Effect (585) of Figure (8-1) since *"Particle Oscillation" is being used as a "Energy Generator"* (EGpo), as so subscribed in WFC

Stanley A. Meyer   8 - 4

RE: Voltage Wave-Guide   Memo WFC 427

memo (424) titled Atomic Energy Balance of Water as to the functional parameters associated with Dynamic Voltage Potential Wave-form (600) of Figure (6-3) ...which uses Voltage Pulse Potential of opposite electrical polarity of attraction (RR' ~ SS' as to RU/RU' ~ ST/DST') to perform work in the following sequence of events in an instant of time: Electrical Polarization Process (160) (elongating the water molecule ... changing the time share rate of the covalent electrons ... switching off the covalent bond by attenuating the electromagnetic fields of the electrical stressed atoms undergoing molecule separation; Universal Energy Priming Stage (500) (particle oscillation as a energy generator by deflecting atomic particles under changing electrical stress); Liquid to Gas Ionization Stage (230) (ejecting electrons from the atomic structure under divergent electrical stress); and Thermal Gas Triggering Stage (E9d) (gas igniting the electrically stress combustible gas atoms farthest from the state of electrical equilibrium) ... triggering Hydrogen Fracturing Process (90) (subcritical-state combustible gases being spark-ignited under Electrical Resonance of Stress).

In terms of Particle Oscillation (Poc) *as* a Energy-Generator (EGpo), if Voltage Arc Line (Val) length is extended while Voltage Amplitude (xx 64a ~ 64b ~ 64c ~ Vn) is adjusted to higher Voltage Peak-Potential (Vpp) then greater atomic interaction (585) of Figure (8-1) (see WFC memo 424 titled Atomic Energy Balance of Water, once again) occurs when particle oscillation (Poc) of deflection of atomic mass (see 550 of Figure 5-8) (atom elongation) is *electrically stressed farthest from the point of state of atomic-equilibrium* by way of opposite Voltage Electrical Attraction Force (RR' ~ SS' as to RU/RU' ~ ST/ST'), as further illustrated in (500) of Figure (5-1) as to (510) of Figure (5-2). Voltage Tickling of State Space under "Resonant Electrical Stress" without amp influxing while "Tuning-In" to the dielectric properties of water is herein referred to in this WFC Tech-manual as "Resonant Action," as illustrated graphically in Figure (5-4 A,B,C) as to Resonant Cavity (170) of Figure (3-25) as to Figure (1-13).

Energy Vectoring (Ev)

The "mode of operability" of determining the "Operational Parameters" of adjusting the thermal explosive energy (gtnt) exiting from nozzle-port (32) of Figure (4-5) as to (40) of Figure (4-2) is directly related to the characteristics of the applied Voltage Pulse Potential (Vpp) Wave-form (s) (Vpwf) and the geometrical configuration of Resonant Cavity (90) of Figure (4-7) as to (730) of Figure (7-12). In terms of Voltage Pulse Wave-form (s) (Vpwf) several "Electrical Operational Parameters" exists: Dynamic State Space (Dss) which continually changes/varies Electrical Attraction Force (RR' ~ SS' as to RU/RU' ~ ST/ST') from low stress intensity (S-low) to high

RE: Voltage Wave-Guide   Memo WFC 427

stress intensity (S- high) and back to low stress point (S-low) as to Arc Curve (Vac) forming Voltage Pulse Field (Vpf) atop Voltage Pulse Burst (Vpb) ...which combined together (Vpf + Vpb) Electrical Stress (Es) variances corresponds to the Voltage Pulse Shape of each synchronized opposite Voltage Pulse Wave (583 ~ 602) of (770A) of Figure (8-1) being produced during applied Voltage Pulse Operation (49a xxx 49n); Static State Space (Sss) is the electrical condition by which Electrical Attraction Force (RR' ~ SS' as to RU/RU' ~ ST/ST') is being held constant once Voltage Pulse Burst Vpb) occurs during Voltage Pulsing Operation (Vpwf) ...forming synchronized Clipped Voltage Wave-form (780C) of Figure (8-2) in like manner to voltage sync-pulse (583 ~ 602).

In the area of Voltage Sync-Wave (+/-) propagation, Unipolar Voltage Pulse Train (583/602a xxx 583/602) of (770A) of Figure (8-1), clipped Voltage Pulse Train (605/606a xxx 605/606n) of Figure (780C) of Figure (8-2), and Crossover Unipolar Pulse Train (607/608a xxx 607/608n) brings-on Static Voltage Stimulation (Vsvs) by which Static Electrical Charging Effect (585) is being held constant since Electrical Stress Force (Esf) averages out either Dynamic State Space (Dss) or Static State Space (Sss) during repeated pulsing operation (49a xxx 49n). On the other hand, Progressive Voltage Sync-Wave (+609 / - 611) (609/611a xxx 609/611n) of Figure (780A) of Figure (8-2) encourages Dynamic Voltage Stimulation (Dvs) since Voltage Peak Potential (Vpp) increases as Voltage Sync-Wave Front (<u>a</u> to <u>b</u> to <u>c</u> and so on) advances in the number of Unipolar Voltage Pulse (s) (Vwp), as illustrated in Figure (3-21) ... causing Dynamic State Space (Dss) or Static State Space (Sss) to be progressively increased in Electrical Stress Intensity (Esi) during a given space-time continuum ...producing Dynamic Electrical Charging Effect (612) of Figure (8-1) that increases Electrical Stress Pressure (Espa + Espb + Espc, and so on) continually during each gated voltage pulsing cycle (49a xxx T3 xxx 49n). To further adjust incoming Voltage Priming Stage (Vps) Unipolar Voltage Pulse Train (Vpt) is either gated full-on to allow space-time continum or back-off in gated format from 100% to a lower percent (%) of Pulse-Frequency on-time, as illustrated in Figure (3-20).

VIC Voltage Sync-Pulse Circuit

Voltage Sync-Pulse Gated Frequency (583 / 602a xxx 583 / 602n) (603/604a xxx 603/604n) of Figure (8-1) as to (605/606a xxx 605/606n) (607/608a xxx 607/608n) (609 / 611a xxx 609 / 611n) of Figure (8-2) ... all, forming Voltage Pulse Burst Wave (619) as to Unipolar Pulse-Train (780A), Crossover Unipolar Pulse-Train (780B), and Clipped Unipolar Pulse Train (780C) as to Traveling Voltage Wave-Action (770) of Figure (8-1) of opposite voltage polarity

RE: Voltage Wave-Guide   Memo WFC 427

(+/-) of equal Voltage-Pulse Amplitudes (+Vpp/-Vpp) are zero reference to electrical ground state (0V) by placing Amp Inhibitor Circuit (860) (Amp Inhibiting Coil 617, Blocking Diode 618, and Magnetic Induction Core 619) between electrical ground (0V) and Center Tap of Dual Bifilar Secondary Pickup Coils (616A/B) of VIC Matrix Circuit (690) of Figure (7-8) as to VIC Impedance Network Circuit (620) of Figure (7-1) , as illustrated in (840) of Figure (8-10). By doing so, Balance Phasing of opposite voltage intensity (+Vpp / - Vpp) is accomplished without experiencing current influxing caused by differential variances where Negative Voltage Peak Potential (-Vpp) is less than Positive Voltage Peak Potential (+Vpp) or Vise Versa ... allowing Inductor Resonant Choke Coils Electromagnetic Fields Intensity (+Z2 / -Z3) to be, in turn, free of Electromagnetic variances of intensity (Z2 ~ Z3). This non-voltage shift (Balance Phasing of opposite Voltage Potential) helps prevents atom displacement during "Snapping-Action" by which "Resonant Electrical Stress" of opposite electrical polarity (RU/RU' ~ ST/ST') is applied equally across Water Molecule (s) (85) to propagate either Static (585) or Dynamic (612) Electrical Charging Effect (s) at elevated Voltage Peak Potential (s). Amp Inhibiting Coil-Assembly (617) is made up of magnetic inductance Stainless Steel 430F/FR wire material wrapped around a closed-loop Induction Magnetic Core (619) which is a separate coil-unit (860) apart from VIC Coil Assembly (580) of Figure (6-1). Blocking Diode (618) functions as an "Electrical Isolator" that prevents electrical discharge of Dual Secondary Coil (616A / B) during applied Pulsing Operations (49a xxx 49n).

    To ensure and maintain Capacitance Charging Effect (650) of Figure (7-4) across Water-Gap (Cp) of (7-8) during applied pulsing operations (49a xxx 49n), Crossover Voltage Wave-Form (780B) as to (780C) of Figure (8-2) is generally utilized by not allowing Convergent Point "Q" of Figure (780B) to reach Electrical Ground Point (0V) when each Unipolar Voltage Pulse (Vpp) is electrical energized in phase-distance relationship to cause the trailing edge (Vpb) of the first Voltage-Pulse (Vpp1) to meet the uprising leading edge (Vpa) of the second Voltage Pulse Wave (Vpp2) at a distance above ground state (0V) determined by the Space-movement of the reforming Voltage Peak Wave (Vppa xxx Vppn) within Voltage Pulse Width (T1), as illustrated in Rotary Crossover Voltage Sync-Pulse Circuit (850) of Figure (8-11) where each VIC Pickup Coils (52A ~ 52B ~52C) are axially spaced 120° apart to cause Convergent Point "Q" to be located 1/3 the height of Voltage Amplitude Peak Level (Vpp), as an example.

Resonant Cavity Configuration (s)
    In reference to Resonant Cavity geometrical configuration to bring-on further response to Energy Vectoring (Ev) apart from Voltage Priming Stage (Vps) (applied incoming Voltage Pulse-Frequency 49a xxx 49n), Resonant Cavity (730) physical design parameter (s) is/are, now, taken

Stanley A. Meyer

RE: Voltage Wave-Guide   Memo WFC 427

into account in determining flame-heat projection (16), as illustrated in (70) of Figure (4-5) as to (590) of Figure (6-2): (A) Linear Cylindrical Resonant Cavity; (B) Taper Cylindrical Resonant Cavity; and (C) Non-Linear Cylindrical Resonant Cavity, as illustrated in (730) of Figure (7-12). Linear Cylindrical Resonant Cavity (730A) Static Variable Condition (Svc) (790) of Figure (8-3) allows Flame-Front (VS1)) of Figure (8-7) to be of equal magnitude when thermal explosive energy-yield (gtnt) is compared with thermal heat-energy (Teh) since Electrical Stress Factor (RR' ~ SS' as to RU/RU' ~ ST/ST') is constant with Static Gas Pressure (Sgp), as illustrated in Energy Vector graph (830) of Figure (8-7). Dynamic Variable (800) of Figure (8-4) produces Flame Front (VS3) when Dynamic Electrical Stress (RR' ~ SS' as to RU/RU' ~ ST/ST') and Dynamic Gas Pressure (Dgp) are both increasing in magnitude by the use of Taper Cylindrical Resonant Cavity (730B) of Figure (7-12), as further indicated in Energy Vector graph (830) of Figure (8-7) under the titled-line "Progressional State Space" (VS3). Combinatorial Variable exists when Dynamic Voltage Pulse Wave (600) of Figure (6-3) is utilized with Linear Cylindrical Resonant Cavity (730A) of Figure (7-12), as illustrated in performance graph (810) of Figure (8-5). Greater Heat-yield of Flame Front (16) is realized when Non-Linear Cylindrical Resonant Cavity (730C) of Figure (7-12) is used in conjunction with Differential Dynamic Variables (820) of Figure (8-6) where Dynamic Electrical Stress (RR' ~ SS' as to RU/RU' ~ ST/ST') is increased while the resultant gas pressure is allow to drop during the gas ignition stage (E9d) of Figure (6-3) as to Figure (590) of Figure (6-2), as finally noted by Energy Vector Graph (830) of Figure (8-7) under titled-line "Expanding State Space" (VS4).

Application of Usage

By simply intermixing/interchanging any applied Electrical Voltage Pulse-State with any Gas Pressure State as herein described above can result in a predetermined hydrogen Gas Flame-Front that can be utilized for a particular application of usage. For example, Taper Resonant Cavity (590) of Figure (6-2) as to (820B) of Figure (8-6) is ideally suited for internal combustion I. C. engines as well as Rocket Engines where high thrust-yield of explosive power (gtnt) (582B) is required; whereas, Expanding Resonant Cavity (730C) of Figure (7-12) as to (820C) of Figure (8-6) is best suited for Furnace Applications. Linear Resonant Cavity (730A) of Figure (7-12) as to Figure (820A) is for Cutting-Torch applications (582) ... to mention a few. In each and all Flame-Front (582A,B,C,) Resonant Pulse Waves are produced to net higher energy-yield beyond normal gas burning levels. Laser Energy (588) being injected into Resonant Pulse Waves (16) by way of Laser Inject Tube-Port (589) helps maintain Plasma-temperatures at extremely elevated temperatures over the prior art. Δ

Stanley A. Meyer

RE: Voltage Wave-Guide   Memo WFC 427

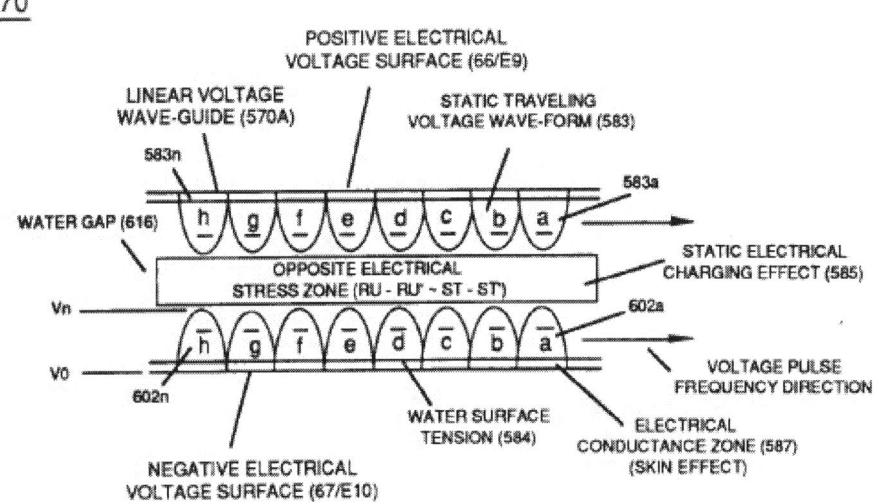

770

(A) STATIC VOLTAGE STIMULATION

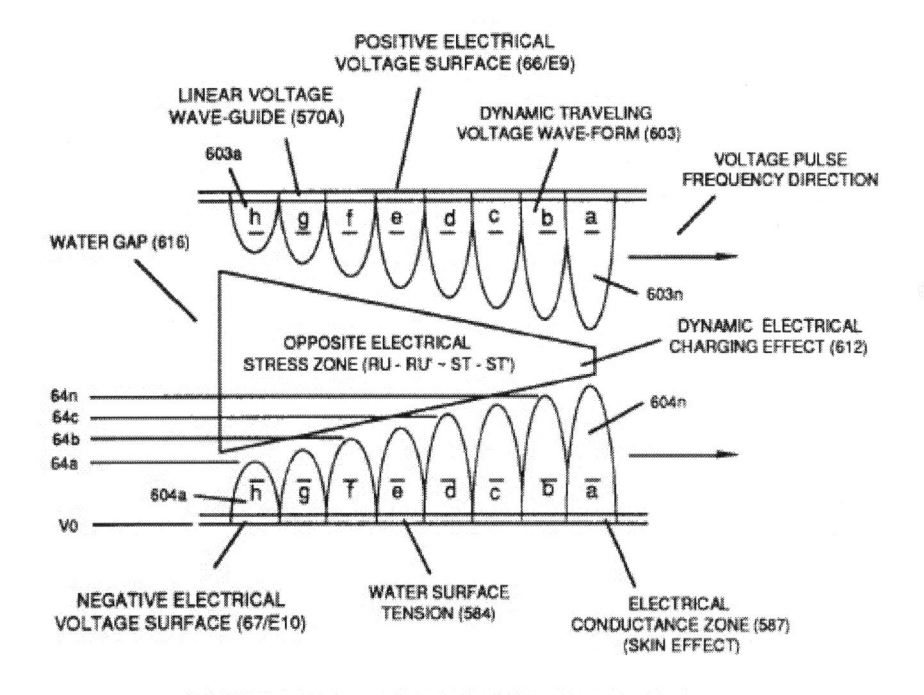

(B) DYNAMIC VOLTAGE STIMULATION

FIGURE 8-1: ELECTRICAL CHARGING STAGE

RE: Voltage Wave-Guide Memo WFC 427

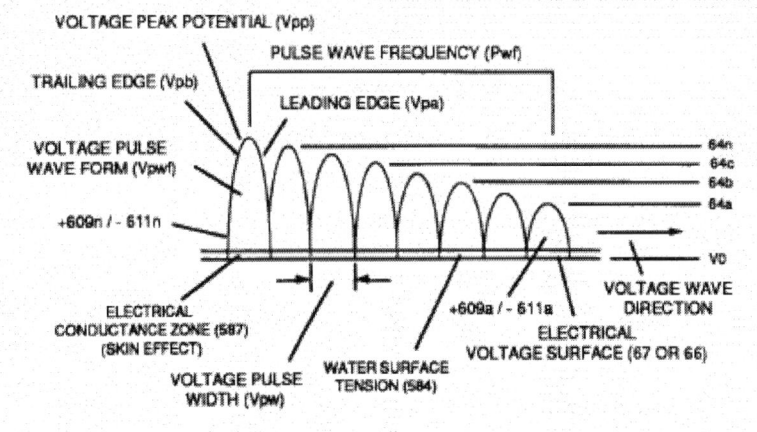

780

Labels (A): VOLTAGE PEAK POTENTIAL (Vpp); PULSE WAVE FREQUENCY (Pwf); TRAILING EDGE (Vpb); LEADING EDGE (Vpa); VOLTAGE PULSE WAVE FORM (Vpwf); +609n / -611n; 64n, 64c, 64b, 64a; Vo; VOLTAGE WAVE DIRECTION; ELECTRICAL CONDUCTANCE ZONE (587) (SKIN EFFECT); VOLTAGE PULSE WIDTH (Vpw); WATER SURFACE TENSION (584); +609a / -611a; ELECTRICAL VOLTAGE SURFACE (67 OR 66)

(A) PROGRAMMABLE UNIPOLAR PULSE-TRAIN

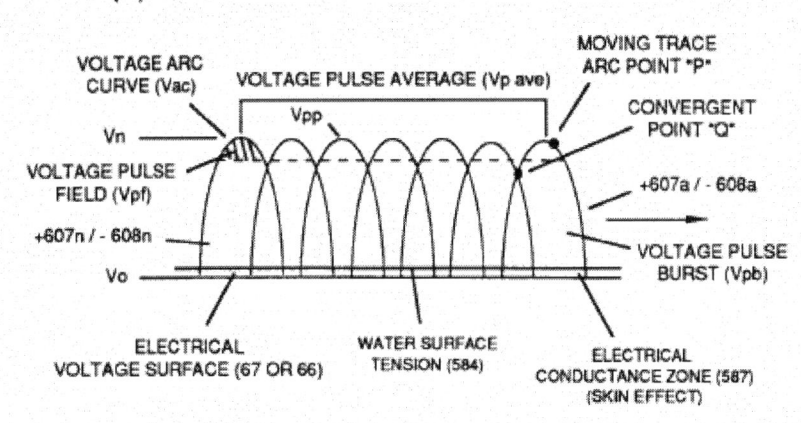

Labels (B): VOLTAGE ARC CURVE (Vac); VOLTAGE PULSE AVERAGE (Vp ave); MOVING TRACE ARC POINT "P"; Vn; Vpp; CONVERGENT POINT "Q"; VOLTAGE PULSE FIELD (Vpf); +607a / -608a; +607n / -608n; Vo; VOLTAGE PULSE BURST (Vpb); ELECTRICAL VOLTAGE SURFACE (67 OR 66); WATER SURFACE TENSION (584); ELECTRICAL CONDUCTANCE ZONE (587) (SKIN EFFECT)

(B) CROSSOVER UNIPOLAR PULSE-TRAIN

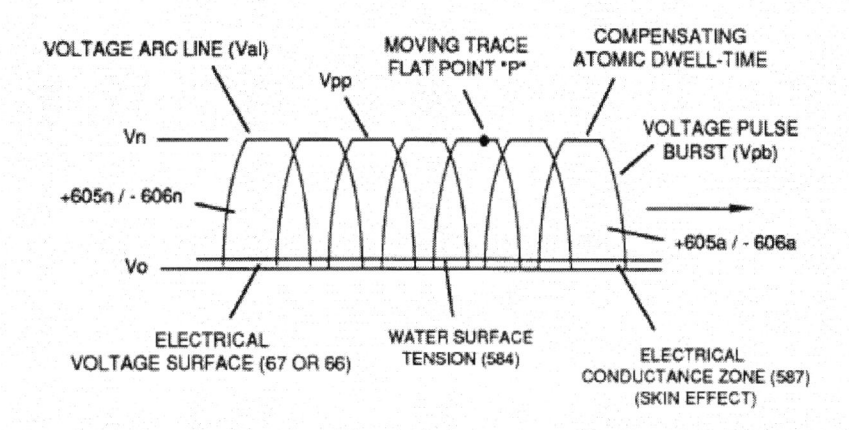

Labels (C): VOLTAGE ARC LINE (Val); MOVING TRACE FLAT POINT "P"; COMPENSATING ATOMIC DWELL-TIME; Vpp; Vn; VOLTAGE PULSE BURST (Vpb); +605n / -606n; +605a / -606a; Vo; ELECTRICAL VOLTAGE SURFACE (67 OR 66); WATER SURFACE TENSION (584); ELECTRICAL CONDUCTANCE ZONE (587) (SKIN EFFECT)

(C) CLIPPED UNIPOLAR PULSE-TRAIN

FIGURE 8-2: PROGRAMMABLE VOLTAGE PULSE-WAVE

Stanley A. Meyer

RE: Voltage Wave-Guide   Memo WFC 427

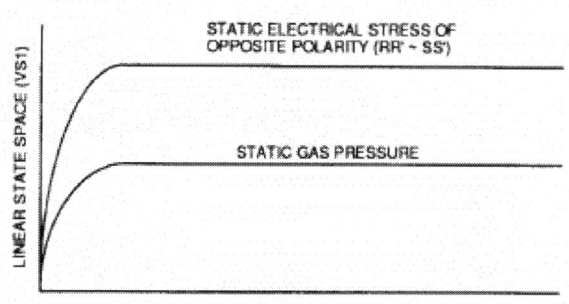

<u>790</u>

FIGURE 8-3: STATIC VARIABLES

(Y-axis: LINEAR STATE SPACE (VS1); X-axis: LINEAR CYLINDRICAL RESONANT CAVITY (730A))
Curves: STATIC ELECTRICAL STRESS OF OPPOSITE POLARITY (RR' ~ SS'); STATIC GAS PRESSURE

<u>800</u>

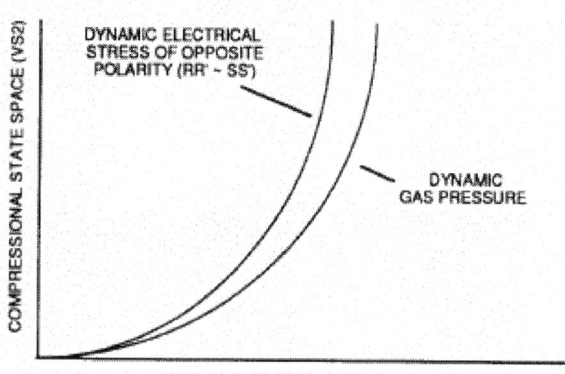

FIGURE 8-4: DYNAMIC VARIABLES

(Y-axis: COMPRESSIONAL STATE SPACE (VS2); X-axis: TAPER CYLINDRICAL RESONANT CAVITY (730B))
Curves: DYNAMIC ELECTRICAL STRESS OF OPPOSITE POLARITY (RR' ~ SS'); DYNAMIC GAS PRESSURE

<u>810</u>

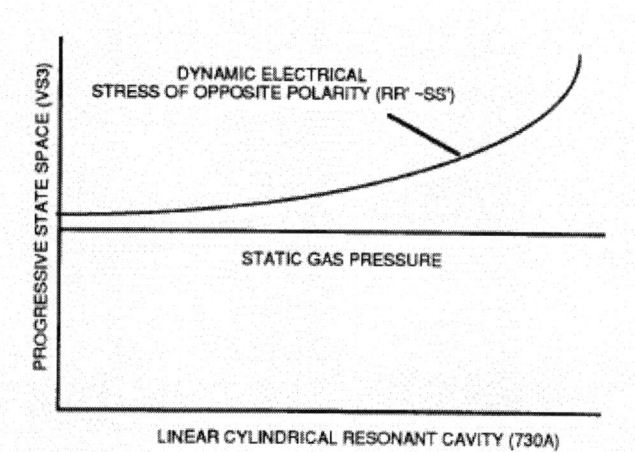

FIGURE 8-5: COMBINATORIAL VARIABLES

(Y-axis: PROGRESSIVE STATE SPACE (VS3); X-axis: LINEAR CYLINDRICAL RESONANT CAVITY (730A))
Curves: DYNAMIC ELECTRICAL STRESS OF OPPOSITE POLARITY (RR' ~SS'); STATIC GAS PRESSURE

RE: Voltage Wave Guide   Memo WFC 427

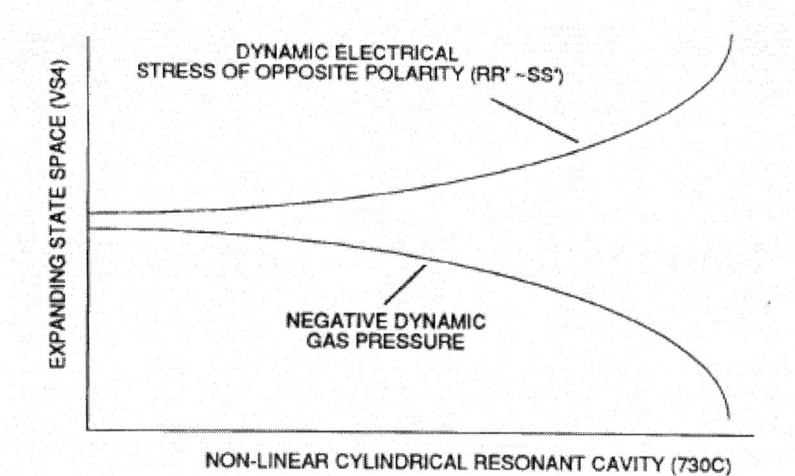

820

FIGURE 8-6: DIFFERENTIAL DYNAMIC VARIABLES

830

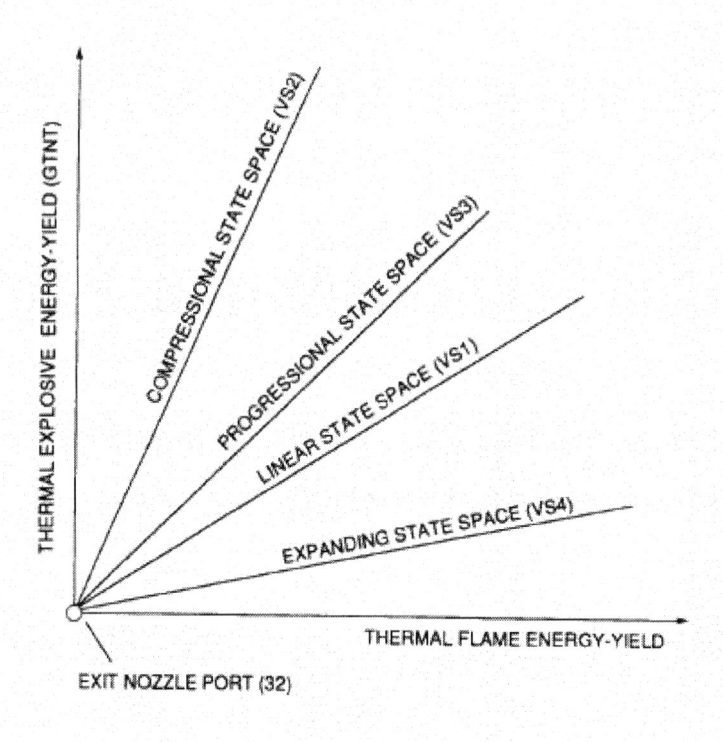

FIGURE 8-7: ENERGY VECTOR GRAPH

RE: Voltage Wave-Guide   Memo WFC 427

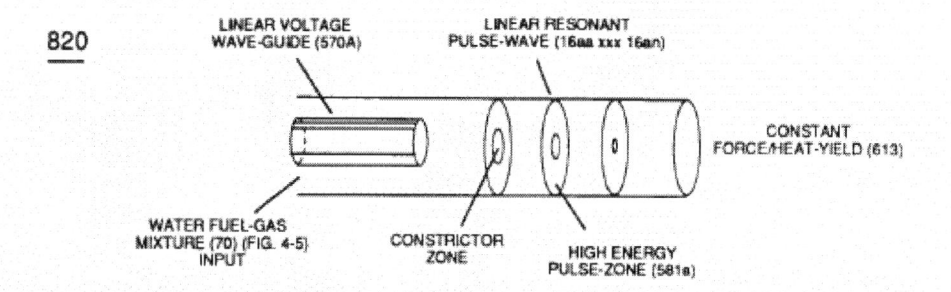

820

(A) LINEAR STATE SPACE (VS1)

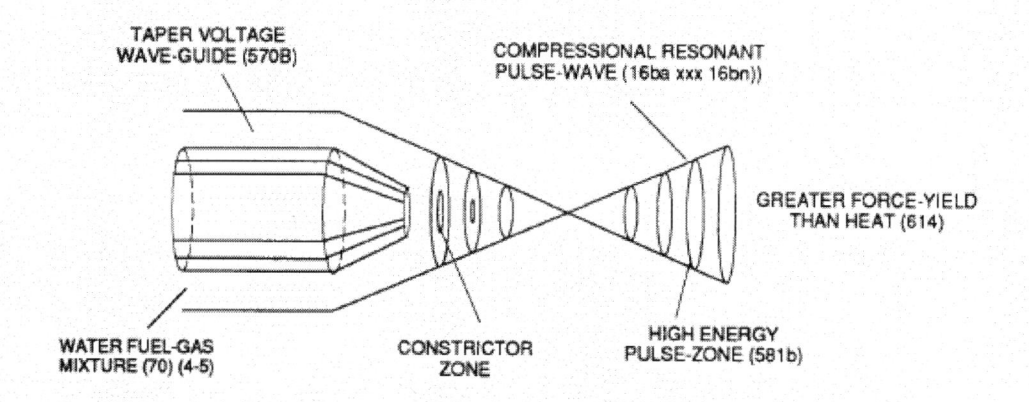

Labels: LINEAR VOLTAGE WAVE-GUIDE (570A); LINEAR RESONANT PULSE-WAVE (16aa xxx 16an); CONSTANT FORCE/HEAT-YIELD (613); WATER FUEL-GAS MIXTURE (70) (FIG. 4-5) INPUT; CONSTRICTOR ZONE; HIGH ENERGY PULSE-ZONE (581a)

(B) COMPRESSIONAL STATE SPACE (VS2)

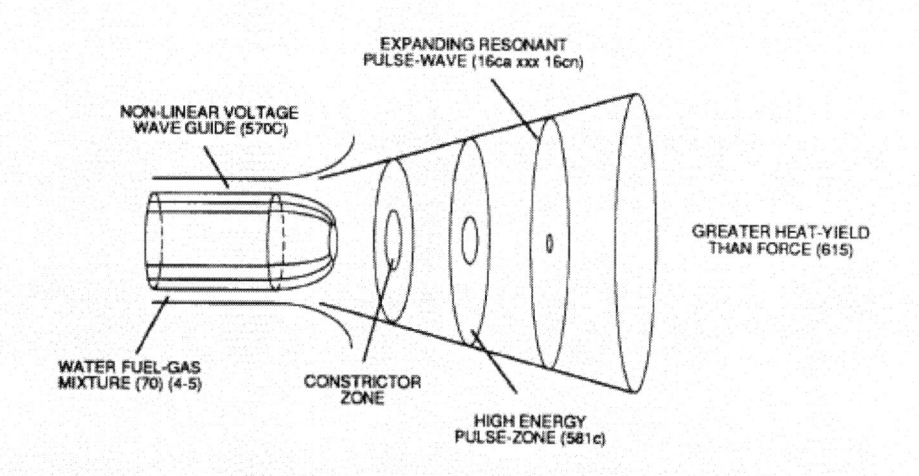

Labels: TAPER VOLTAGE WAVE-GUIDE (570B); COMPRESSIONAL RESONANT PULSE-WAVE (16ba xxx 16bn); GREATER FORCE-YIELD THAN HEAT (614); WATER FUEL-GAS MIXTURE (70) (4-5); CONSTRICTOR ZONE; HIGH ENERGY PULSE-ZONE (581b)

(C) EXPANDING STATE SPACE (VS4)

Labels: NON-LINEAR VOLTAGE WAVE GUIDE (570C); EXPANDING RESONANT PULSE-WAVE (16ca xxx 16cn); GREATER HEAT-YIELD THAN FORCE (615); WATER FUEL-GAS MIXTURE (70) (4-5); CONSTRICTOR ZONE; HIGH ENERGY PULSE-ZONE (581c)

FIGURE 8-6: RESONANT CAVITY STATE SPACE

Stanley A. Meyer   8-13

RE: Voltage Wave-Guide   Memo WFC 427

830

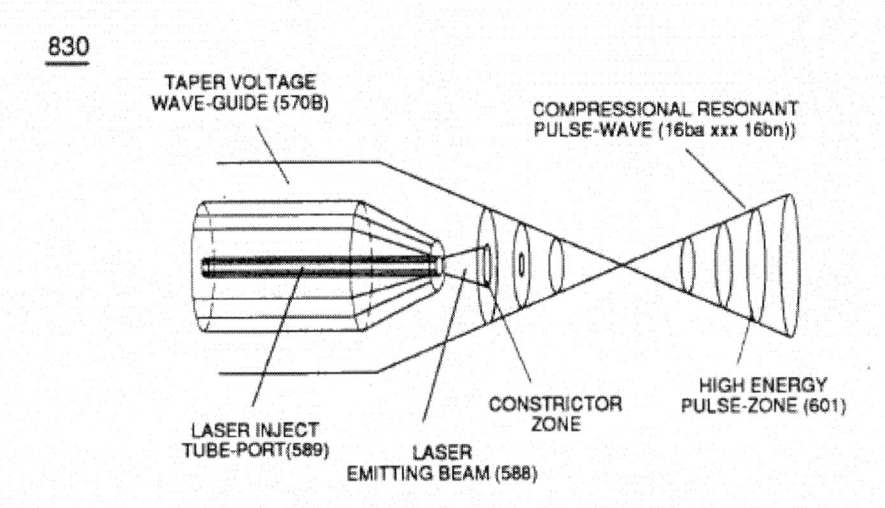

FIGURE 8-9: LASER ENERGY INTERACTION

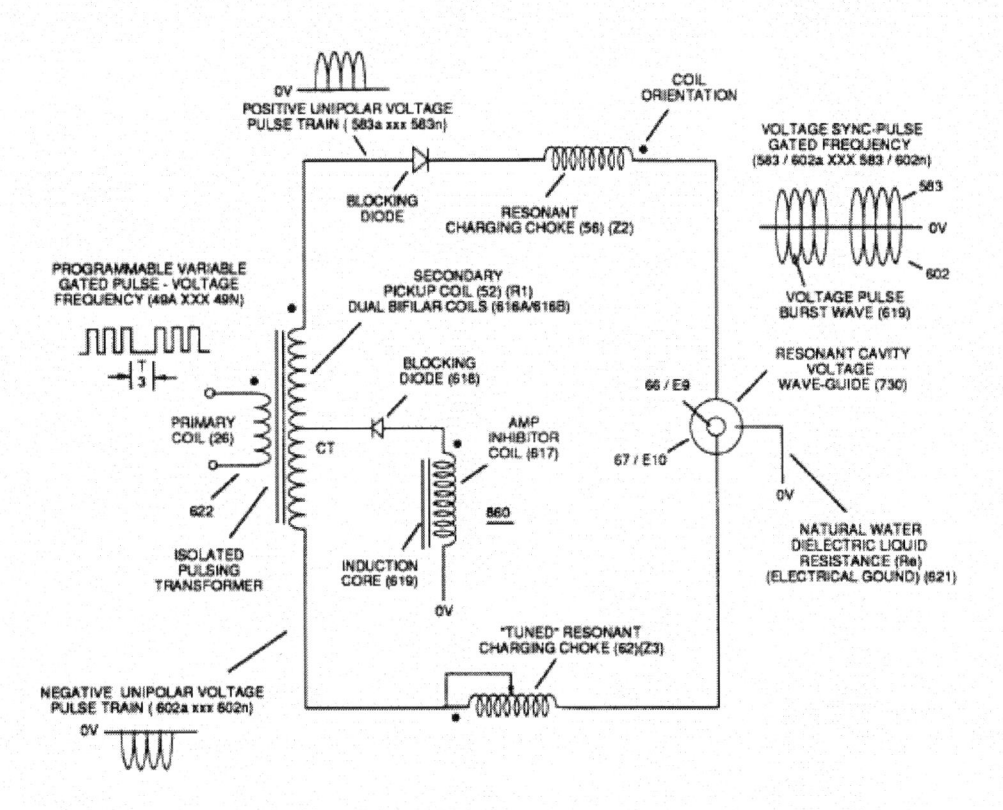

840

FIGURE 8-10: VIC VOLTAGE SYNC-PULSE CIRCUIT

Stanley A. Meyer   8-14

RE: Voltage Wave-Guide   Memo WFC 427

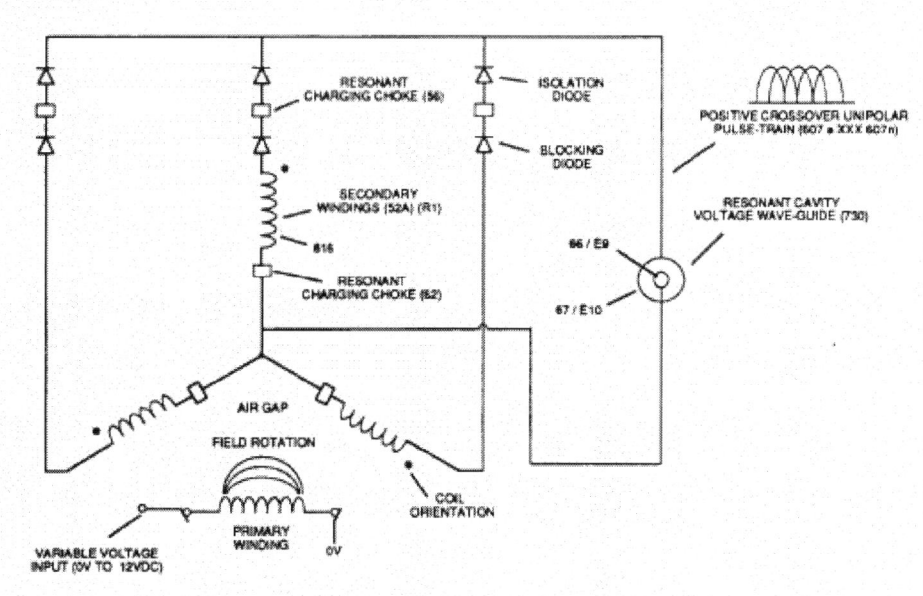

(A) UNIPOLAR CROSSOVER VOLTAGE PULSE-TRAIN

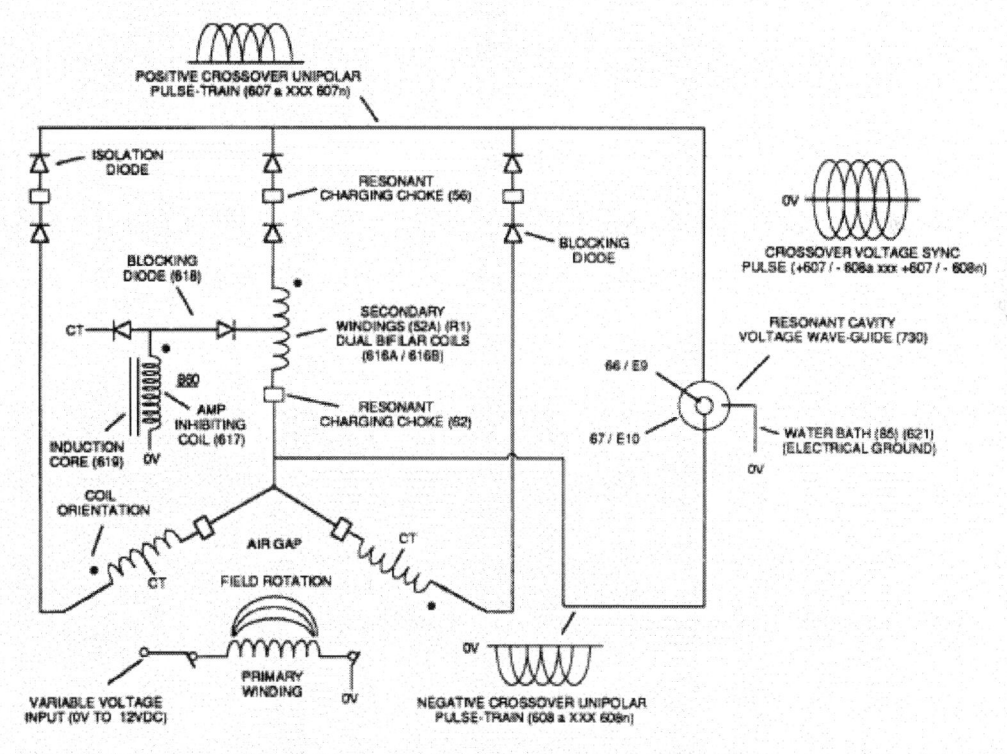

(B) CROSSOVER VOLTAGE BURST SYNC-PULSE

FIGURE 8-11: ROTARY (VIC) PULSE VOLTAGE FREQUENCY GENERATOR

Memo WFC 428

Reclaiming Our Air ... For Healthy Living

The prime objective of the Water Fuel Injection Technology of Invention (s) is to help reverse the damage being done to "Earth Ecological Life Support System" by first of all encouraging the use of "Water" as a new "Fuel" source since the by-product of releasing thermal explosive energy (gtnt) from water is simply "Water Mist" which is energy recyclable by absorbing solar-energy from the incoming sun rays... automatically limiting the use of fossil-fuel burning ... stopping the extraction of Oxygen O_2 molecules from the air since "Water" supplies it own oxygen molecule to support the hydrogen combustion process.

Secondly, re-energizing the energy-level of the air by the use of WFC Gas Processor by tapping into "Universal" energy by way of particle oscillation as a energy generator.

And lastly, make use of WFC Exhaust Air Reclaiming technology to unlock and do away with airborne chemical-oxides derived from fossil-fuel burning when ambient air passes through the internal combustion (IC) engine running on water ...thereby repurifying our air for healthy living while maintaining the Industrial economy of the World. Remember, water, of course, is free, abundant, and energy recyclable.

WFC Exhaust Air Reclaimer technology utilizes the WFC Electrical Polarization Process to reverse the degradation effect of burning fossil-fuel ...liberating chemical-oxides oxygen O_2 molecules once again for reuse under "Earth Ecological Life Support System" ...collecting atoms under static charged for industrial reuse ...resulting in revitalized "Clean" air once again to sustain all life-forms on Earth. WFC Exhaust Air Reclaimer functions in the following way:

Section 9

RE: Exhaust Air Reclaimer   Memo WFC 428

Reclaiming Our Air ... For Healthy Living

In almost all cases of scientific accomplishments, spin-off technology occurs. The Water Fuel Cell technology of Inventions is no exception to this rule. For every cause there is a reason and for every reason there is an answer and for every answer there is progress. Without scientific progress, we can not hope to cope with, nor solve, the needs of the World. The answers given below are for this purpose and this purpose only.

Electrovalent Bonding

In similar manner by which polar Water Molecule unlike atoms (Hydrogen Atoms 78 / Oxygen Atom 81) (210) of Figure (3-27) take-on opposite electrical Charges (B+ / B-), other gas-atoms molecule (s) experience the same Electrical Charge Effect (q - q') when covalent-electron sharing occurs, as illustrated in polar-molecule Carbon Dioxide CO_2 (910) of Figure (9-2) as to allotropic molecule of Ozone O_3 (930) of Figure (9-4). Carbon Dioxide molecule CO_2 (910) Electrovalent Bonding forces (q - q') comes into existence when unlike Carbon Atom (902) shares electrons with each of two Oxygen Atoms (901a / 901b) since the accepted and captured covalent electrons migrates toward both oxygen atom (901a and 901b) nucleus proton-cluster of eight particles having a greater total positive static charge than Carbon Atom (902) nucleus proton-cluster of only six ... forming polar charged (B+ / B-) Carbon Dioxide CO_2 molecule (910). The additive two captured/accepted electrons (total ten 10 electrons as to only eight 8 protons) causes both oxygen atoms (901a / 901b) to individually take-on a negative electrical charge (B-) while the center positioned Carbon Atom (902) emanates a positive electrical charge (B+) of equal but opposite electrical intensity (q - q') when its shared electrons is/are being used/accepted by unlike oxygen atoms (901a / 901b). Nitrogen Dioxide NO_2 (940) of Figure (9-5) is another example of polar electrical charging (q - q') of two unlike atoms forming a stable molecule wherein a Nitrogen Atom (N) (903) covalently interlocks with two Oxygen Atoms (904a /904b). Identical gas-atoms of Oxygen Atoms (905a / 905b / 905c) of Figure (9-4) further exhibits Electrical Charging Effect (q - q') since in all cases the second Electron Shell (L- orbit) can accept up to eight (8) electrons to cause molecule stabilization. Transitional gas-molecule of Oxygen O_2 combines together two oxygen atoms (906a / 906b) in this way while allowing the donor oxygen atom (906b) to except another oxygen atom (905c) of Figure (9-4) since it's L-Orbit (Outer Shell) still has available two "Electron-Gaps" for covalent Linkup, as illustrated in (920) of Figure (9-3).

Electrical Charging Effect (q -q') is Electrical Attraction Force (q - q') of opposite electrical polarity between the established positive (B+) electrically charged atom (s) and the negative (B-) electrical charged atom (s). Electrical intensity of Opposite Electrical Attraction Force (q-qa'- 907a + q-qb' - 907b) (herein after called Electrovalent Bonding) (total electrical bonding force between two opposite electrical charged atoms) are equivalent to the total number of electrons being used/accepted by the host atom (s) having the

Stanley A. Meyer   9 - 1

RE: Exhaust Air Reclaimer   Memo WFC 428

greater positive charged (B+) nucleus as so established under the laws of physics which states for "every action there is an equal and opposite reaction". This is possible due the fact that all orbiting individual electrons display their own negative electrical charge (B-) whereas each proton-particle separately supports a positive electrical charge (B+) ... both opposite electrical charged particles (Proton as to each Electron) being equal in electrical magnitude (B+ =B-). And due to the fact that the oxygen atom does not take-on an electromagnetic charged field since its electrons pair together and spin in opposite direction.

Ecological Support Cycle Changing

Electrovalent Bonding of similar and non-similar atoms to form gas-molecule (s) produces different elements of substance. For example, in the combustion process of burning fossil-fuel (oil, gasoline, diesel-fuel, coal, or natural gas) in air results in unwanted airborne pollutants such as Carbon Monoxide CO, Nitric Oxide NO, Nitrogen Dioxide NO_2, Carbon Monoxide CO, Carbon Dioxide $2CO_2$, Sulfur Dioxide SO_2, and unburned hydrocarbons ... to mention a few. Once these chemical-oxides enters/discharged into the atmosphere surrounding Earth (see graph 950 of Figure 9-6) other chemical interactions (oxidization) may occur under the process of "Photochemical Bonding" by sunlight. For example, when Sulfur Dioxide SO_2 molecule Electrovalent link up to an Oxygen O_2 molecule (920) of Figure 9-3)... forming Sulfur Trioxides $2SO_3$ which when mixed with rain produces "Sulfuric Acid" known Worldwide as "Acid Rain". Toxicity of these air-borne (dilution in air) chemical-oxides (907) of Figure (9-6) is/are not only endangering "Earth Ecological Support System" but in the formation of these airborne chemical-oxides by way of catalytic-action of combustion (burning) are extracting/consuming enormous amounts of free-floating Oxygen O_2 molecule (920) of Figure (9-3) out of the air (less than 20% now as compared to 30% prior to the Industrial Age as so determined by polar ice-cap samples) which is vital to sustain all life-forms on Earth (see Blue Planet cinematography-film by Imax-NASA.).

These increases of airborne pollutants over the years are, also, a major contributor of decreasing Sun-Energy (photon absorption know as the "Green-House Effect") (estimated 10% reduction in some parts of the World today) which is, now, allowing the viro-germ of "Blight" to form ... killing-off forest region / plant -life around the World. Remember, plant-life is required to convert Carbon Dioxide CO_2 (910) of Figure (9-2) into liberated Oxygen O_2 so that the "Breath of Life" is provided for all animal-life. Entrapping the Oxygen O_2 molecule in the chemical-oxide process of burning fossil-fuel (s) is, also, disrupting/preventing the "External Respiration Process" of plants from exchanging oxygen O_2 and Carbon Dioxide $2CO_2$ between the plant-organism and its environment in the present of sun light. The reduction of sun light is further stunting the growth of plants while, simultaneously, causing plant respiration process cycling-rate to decline exponentially. Worst yet, chemical-oxide airborne pollutants derived from burning fossil-fuels are chemically interacting with the Ozone layer 20 miles above Earth ... causing ozone layer depletion... allowing increased ultraviolet radiation originating from the sun to pass through the air troposphere ... contributing not only to changing/altering man's "Ecological Life Support System" on Earth but allowing a portion of Earth "Air Gases" to escape into outer space. Furthermore, free-floating

Stanley A. Meyer

RE: Exhaust Air Reclaimer   Memo WFC 428

chemical-oxides are "Electrovalent-Bonding" to Earth ionosphere ... converting Earth Air Supply into a "Jelly-Like " substance ... entrapping moisture from air ... disrupting Earth Rain Cycle, as so illustrated in (960) of Figure (9-7). Dispersion of radioactive/chemical biological germ-agent (s) contaminated oil supplies coming out of the Arab Gulf War may pose another real threat to man's survival. Inhaling radioactive exhaust fumes from contaminated Arab oil certainly would cause Lung cancer if such a deplorable situation would be allowed to exist even for a brief period of time.

WFC Development Objectives

The primary purpose/objective of Water Fuel Injection System (590) of Figure (6-2) as to "Full-System" development (10) of Figure (4-1) is, in the realm of scientific quest, to help reverse the damage being done to "Earth Ecological Life Support System" by first of all encouraging the use of "Water" as a new "Fuel" source since the by-product of releasing thermal explosive energy (gtnt) from water is simply "Water Mist" which is "Energy Recyclable" by absorbing "Solar-Energy" from the incoming sun rays, as so illustrated in Energy Recycling Spectrum (530) of Figure (5-6) ... automatically limiting the use of fossil-fuel burning... stopping the extraction of Oxygen O_2 molecule from the air since "Water" supplies its own oxygen molecule to support the hydrogen combustion process. Secondly, re-energizing the energy-level of the air by the use of WFC Gas Processor (80) of Figure (1-17) by tapping into "Universal" energy by way of particle oscillation as a energy generator, as so illustrated in Energy Pumping Action (520) of Figure (5-3) as to Energy Aperture (570) of Figure (5-10). And lastly, make use of WFC Exhaust Air Reclaiming technology (900) of Figure (9-1) to unlock and do away with the airborne chemical-oxides derived from fossil-fuel burning when ambient air passes through the Internal combustion (IC) engine running on water ... thereby repurifying our air for healthy living while maintaining the Industrial economy of the World. Remember, Water, of course, is free, abundant, and energy recyclable.

WFC Exhaust Air Reclaimer

To eliminate the possibility of expelling out any Nitric Oxide N0 gas which is generally produced when an electrical spark occurs inside an internal combustion (IC) engine or any other chemical-oxide that may be present in the exhaust-air, WFC Exhaust Air Reclaimer (900) of Figure (9-1) can be utilized to comply with U.S. Environmental Protection Agency (EPA) Clean Air Act. Similar in construction to Voltage Intensifier Circuit (110) of Figure (4-9), WFC Exhaust Air Reclaimer (900) of Figure (9-1) utilizes opposite electrical attraction force (RR' ~ SS') of voltage potential (Va xxx Vn) to separate the atoms of the Nitric Oxide N0 molecule (typically $\leq$ 1ppm when running on water) by overcoming Electrovalent Attraction Force (q - q') that exist between the polar-charged Nitrogen (B+) and oxygen (B-) atoms. This is accomplished since the Voltage Intensifier Circuit (110) of Figure (4-9) is pulsed sensitive (49a xxx 49n) to tune-in to the dielectric properties of the airborne gas-molecule to not only activate Covalent Switch-Off phenomenon (550) of Figure (5-8) by restricting amp influxing to cause "Electrical Stress" across

Stanley A. Meyer   9 - 3

RE: Exhaust Air Reclaimer   Memo WFC 428

capacitor gap (Cp) (690) of Figure (7-6) but, also, elongates the electrically polarized Nitric Oxide N0 molecule to change the time share-rate of covalent electron (s) ... performing the same function as Electrical Polarization Process (160) of Figure (3-26) by which the positive electrical voltage potential (71) attracts the negative charged oxygen atom; while, simultaneously, negative electrical voltage potential (61) attracts the positive charged Nitrogen Atom ... pulling apart and separating the atoms of the Nitric Oxide N0 molecule by way of Voltage Dynamics (220) of Figure (3-29) under the law of physics which states "opposite electrical charges attracts", as so illustrated in (900) of Figure (9-1). Likewise, other chemical-oxide gas-molecule (s) such as Nitrogen Dioxide N02 (940) of Figure (9-5) exposed to Electrical Polarization Process (160) utilizing Gas Resonant Cavity (140) undergoes the same molecule-to-atom separation ...thereby, reversing the degradation effect of burning fossil-fuels ... liberating oxygen 02 molecule once again for reuse under "Earth Ecological Life Support System" ... collecting atoms under static charged for Industrial reuse ... resulting in revitalized clean air once again to sustain all life-forms on Earth. Of course, water mist derived from hydrogen combustion is removed prior to engine exhaust air-gases entering into and passing through WFC Exhaust Air Reclaimer System (900).

Air purification by atomic deflection (900) of Figure (9-1) is possible since airborne atoms are neither destroyed nor created during exposer to "Voltage" pulse-stimulation (E11/E12 ... 49a xxx 49n) propagating "Electrical Stress" pulse-waves ($\Delta RR' \sim \Delta SS'$) which, in turn, induces "Electrovalent Switch-off" phenomenon (550) of Figure (5-8). Energy enhancement of these liberated airborne atoms during the process of molecular-separation (900) comes either from Photon Absorption Process (537) of Figure (5-6), or from Energy Pumping Action (520) of Figure (5-3), or combination of both Energy Sources ... Sun Energy (534) in retrospect to Universal Energy Pathway 9 of Figure 5-10), as so illustrated in "Energy Recycling Spectrum (530) of Figure (5-6). In alternate form, solid state electrical photon-light source (Light Emitting Diodes) (116a xxx 116n) of Figure (3-33) replaces Sun Rays (539) of Figure (5-6) when "Voltage Tickling of State Space" (770 of Figure (8-1) acts as a "energy generator" by way of "particle oscillation" (500 of Figure 5-1) as to (510 of Figure 5-2) to cause/activate Atomic Energy Level Adjustment (540) of Figure (5-7), as illustrated in Gas Processor Process (260) of Figure (3-33) as to Figure (1-17).

Ecological Plus Factors

Other application (s) of the WFC Air Reclaimer technology (900) of Figure (9-1) is/are realized, such as but not limited to: destroying bacteria in human waste-slurry for natural fertilization, omission of many toxic-waste chemicals by atomic separation of there liquid to gas molecules for industrial reuse, and even for static charging sea water contaminates for fresh water supply ...to mention a few. Δ

Stanley A. Meyer

RE: Exhaust Air Reclaimer Memo WFC 428

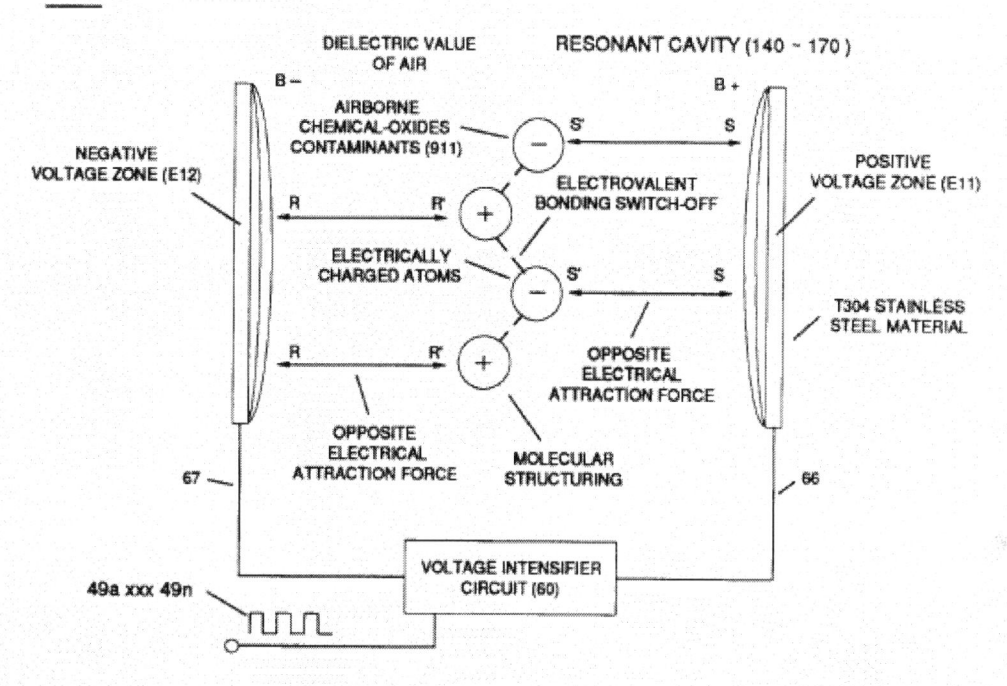

FIGURE 9-1: EXHAUST AIR RECLAIMER

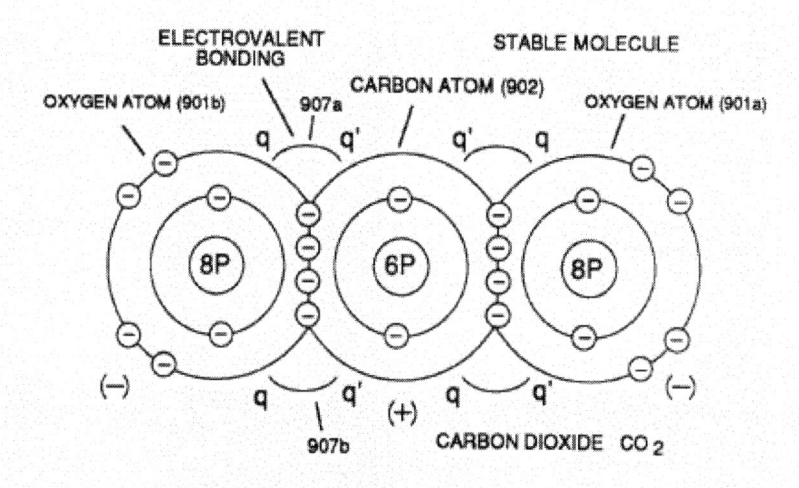

FIGURE 9-2: CARBON DIOXIDE CO_2

RE: Exhaust Air Reclaimer   Memo WFC 428

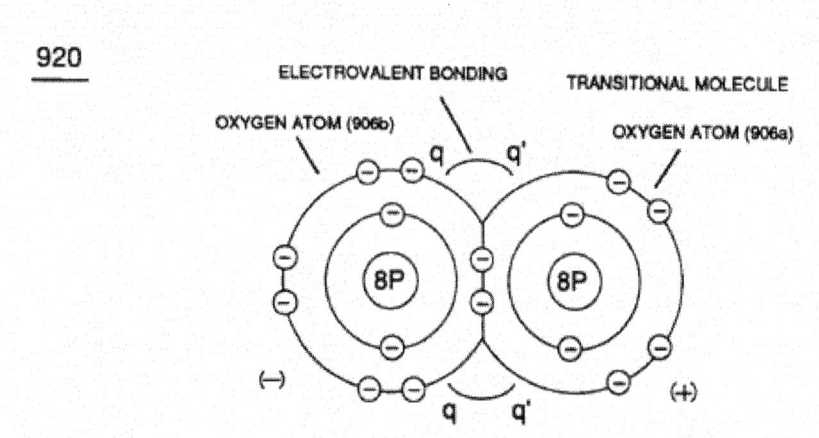

920

FIGURE 9-3: OXYGEN O_2

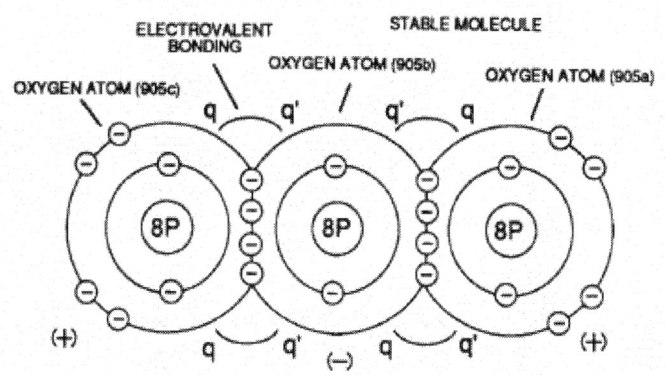

930

FIGURE 9-4: OZONE O_3

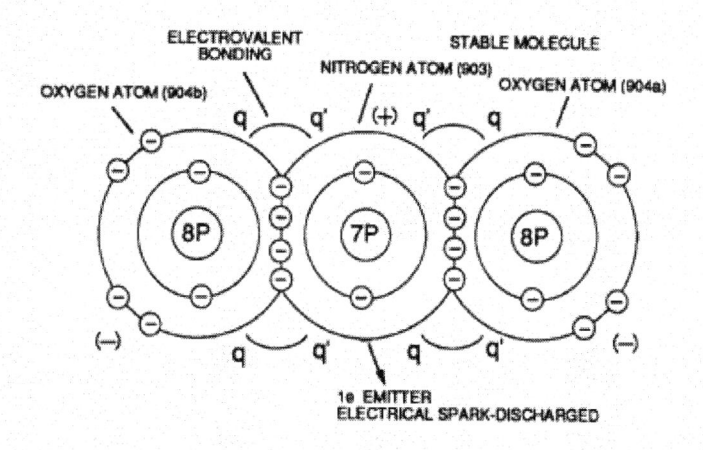

940

FIGURE 9-5: NITROGEN DIOXIDE NO_2

RE: Exhaust Air Reclaimer Memo WFC 428

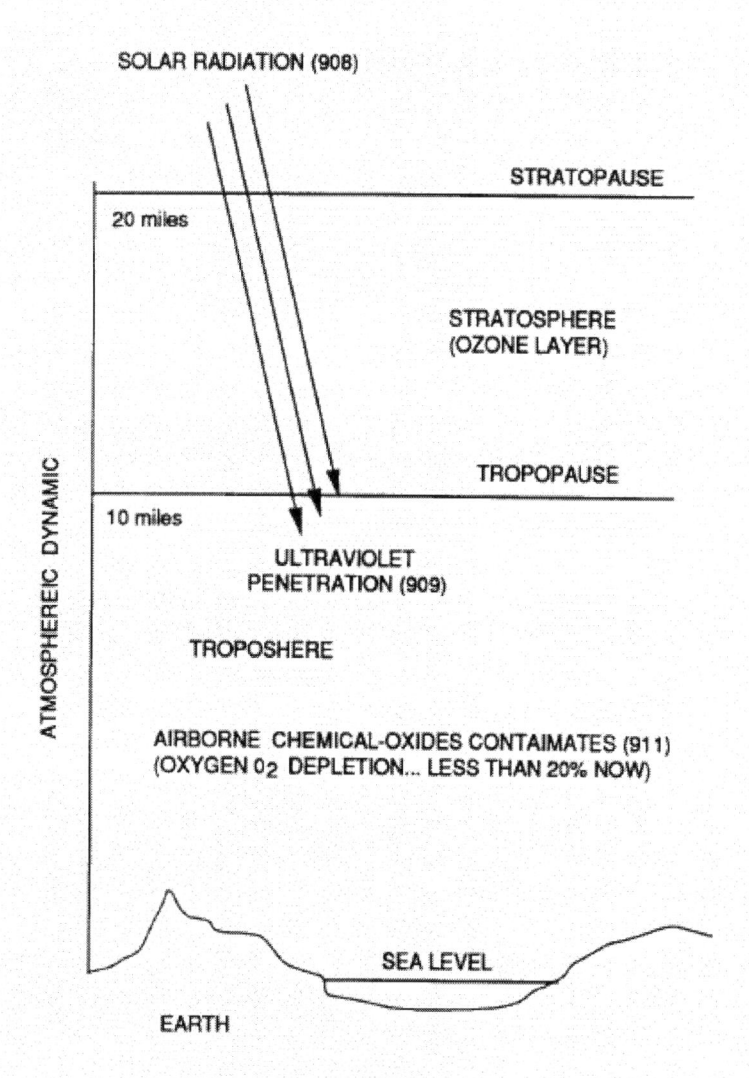

950

```
                SOLAR RADIATION (908)
                        \  \  \
    ─────────────────────\──\──\──────── STRATOPAUSE
    20 miles              \  \  \
                           \  \  \
                            \  \  \       STRATOSPHERE
                             \  \  \      (OZONE LAYER)
                              \  \  \
    ──────────────────────────▼──▼──▼──── TROPOPAUSE
    10 miles
                         ULTRAVIOLET
                         PENETRATION (909)

                         TROPOSHERE

              AIRBORNE CHEMICAL-OXIDES CONTAIMATES (911)
              (OXYGEN O2 DEPLETION... LESS THAN 20% NOW)
```
ATMOSPHEREIC DYNAMIC

SEA LEVEL

EARTH

FIGURE 9-6: RE-PURIFYING EARTH AIR SUPPLY

Stanley A. Meyer 9-7

RE: EXHAUST AIR RECLAIMER   Memo WFC 428

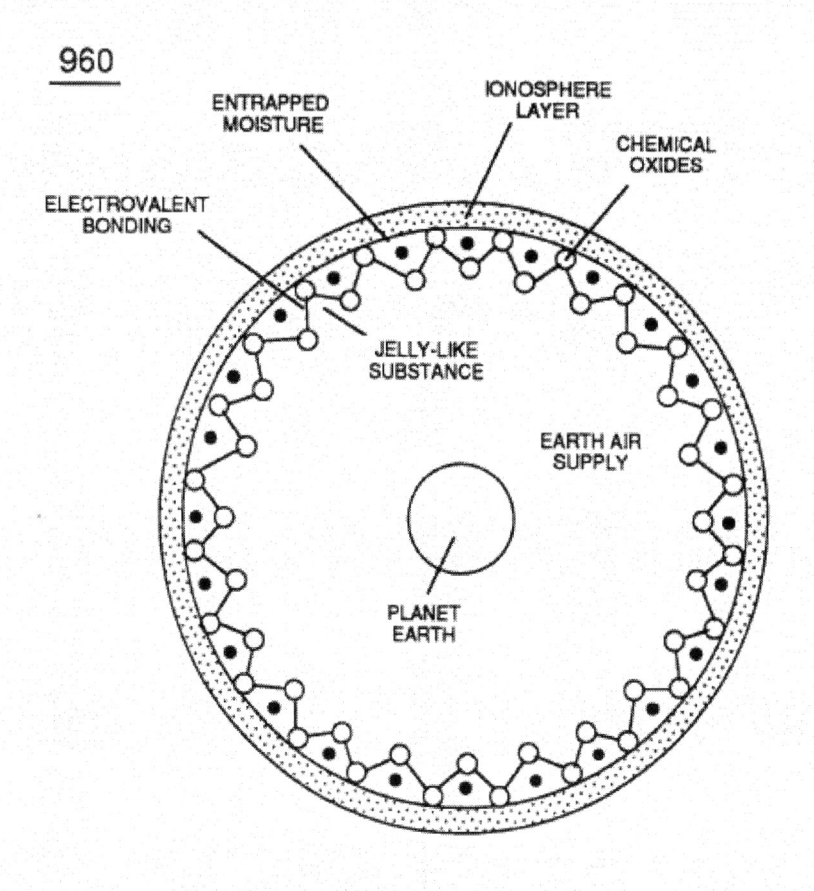

960

FIGURE 9-7: SPACESHIP EARTH

Memo WFC 429

Optical Thermal Lens:

A Miniature Controllable Sun

How is it that a seed of a plant can produce atoms, undergo molecule structuring, support chemical processes that sustains life? The sun, soil nutrients, and water are only growth stimulators.

How is it that that a baby in the mother womb undergoes atomic structuring to give us life? The child mother does not consume enough food nor absorb enough sun rays to create not even one atom of the new born.

How is it mass is being created in our universe? And why is it that planets continue to move in orbital pathways? What fuels the universe? The universe is still expanding while being maintained.

Where does fire come from since atoms are neither created nor destroyed during the burning process?

Why is lighting associated with water molecules in air? And what fuels hurricanes ... is it the Water Molecule?

Einstein $E=MC^2$ equation shows that the energy comes from somewhere beyond our physical universe.

Is "Particle Oscillation" as a "Energy Generator" the way to tap into this universal energy source to fuel our economy? And if so, by what means?

The WFC Technique of "Easer" is, now, open for scientific dissertation ... and is explained in the following way:

Section 10

RE: Optical Thermal Lens   Memo WFC 429

Optical Thermal Lens

A Miniature Controllable Sun

In the annals of scientific endeavors, the prior art only subscribed in using the following scenarios to cause "Particle Oscillation" as a "Energy Generator": particle impact to produce kinetic energy; current flow to manifest light energy by which we see from an incandescent light bulbs; electromagnetic flux-energy to ionize gases inside an fluorescent tube exposed to an electrical power line; photon energy absorption to activate a laser device by the use of a strobe light; chemical stress to cause combustion by rearranging molecular structures; redirecting and splitting light transmission through optical prismatic structures, and the atomic absorption and re-radiation of acoustical energy for sound-box amplification. In reference to WFC Technology of Inventions, "Electrical Stress" propagated by "Opposite Voltage Polarity" is utilized as a "New" way to cause "Particle Oscillation" as a "Energy Generator"... giving way to "the Birth of New Technology" ... such as, "The Optical Thermal Lens" called "Easer"... tapping into "Universal Energy" via "Voltage Tickling of State Space" under resonant conditions.

Propagating Electrical Stress

As in reference to WFC Patent Validation Report dated January 14, 1983 as per WFC Test-Results "Mode of Operability" of using "Voltage Potential" to "Dissociates the Water Molecule" by way of the "Electrical Polarization Process"(160) of Figure (3-26) as so specified under U.S. Patent Law (35 USC 101) to demonstrate operability, the applied Pulse-Voltage Frequency is adjusted to tune-in to the dielectric properties of water by the use of WFC "Amp Inhibiting Circuit" (970) of Figure (10-1), as further illustrated in WFC Tech-Brief titled "The Birth of New Technology"... U.S. Patent Memos 420 ~ 428, including "Table of Tabulation" (Appendix A) as to "Glossary of Application Notes" (Appendix B).

The Amp Inhibiting Circuit (970) of Figure (10-1) as to (690) of Figure (7-8) is composed of two copper wires "Bifilar" wound (wrapped) about a magnetic induction core to allow amp restriction (minimizing current leakage) while encouraging "Voltage Potential"(Va xxx Vn) across the water molecule to perform WFC "Electrical Polarization Process", as so illustrated in Figure (7-1) WFC memo (426) titled VIC Matrix Circuit. The energized "Resonant Charging Choke" (56) of Figure (7-1) as to Figure (10-1) by way of input voltage-pulses (49a xxx 49n) creates an electromagnetic coupling field (Rp1) of Figure (7-8) due to its self-inductance (640) of Figure 7-3B) crosses over and passes through electrically ground connected Resonant Charging Choke (62), as so illustrated in Figure (10-1) ... causing amp flow restriction during each pulsing-cycle since electrons exhibit electromagnetic characteristic ... forming "Mutual Inductance Fields" (Rp1/Rp2) once secondary coil (62) is electromagnetically energized by primary coil (56) and vice

Stanley A. Meyer

RE: Optical Thermal Lens   Memo WFC 429

versa ... thereby, preventing amp "in-fluxing" (discouraging electron arc over) across Dielectric Capacitor Gap (ER)(66/67) while Electrical Stress (ST-ST' ~ RU-RU') of Opposite Voltage Polarity (B+/B-) brings on Energy Priming Stage (520) of Figure (5-3) which is refer to, herein, as "Voltage Tickling of State Space."

The resultant Amp Inhibiting Circuit Figure (10-1) as to Figure (10-3A/B) further allows amp restriction (minimizing current leakage) to be continued even if applied "Voltage Amplitude" is increased. The length and diameter size of the copper-wire spiral wrapped coil (56/62) of Figure (10-1) being paired together and electrically energized in conjunction with applied Voltage Pulse-Frequency determines how much "Amp Leakage" will occur across capacitor Gap (Cp) while "Voltage Pulse-Potential" (Va xxx Vn/49a xxx 49n) of "Opposite polarity" (B+/B-) is/are allowed to be applied across "Electrical Voltage Plates " (Voltage-Zones) (66/67). To reduce amp leakage still further, the copper wire of both Resonant Charging Chokes (56/62) can be replaced with an magnetically inductive stainless steel wire (430F/FR) having a resistive value (Ohms) to the flow of electrons while taking on the capacitance and inductance characteristic of a coil wire. VIC Bifilar Wrap Coil-Assembly (10-3B) and VIC Dual Coil Wrap-Assembly (10-3A) both utilize either "E"& "I" and "U" Inductance Core configurations to concentrate Mutual Inductance Fields (Rp1/Rp2) in order to optimize Amp Inhibiting Process (750) of Figure (7-14). "E" & "I" core shape (10-3B) is most preferable since amp spike surge is minimize during repetitive pulsing operations.

Beyond amp restricting characteristic of said Amp Inhibiting Circuit (970) of Figure (10-1) as to Voltage Intensifier Circuit (60) of Figure (3-22), the spiral-wrapped coils (Resonant Charging Chokes 56/62) being paired together, also, causes voltage level enhancement beyond applied voltage input since the "Distributed Capacitance" (C1a xxx C1n ~ C2a xxx C2n) / "Distributed Inductance" (FL1a xxx FL1n ~ FL2a xxx FL2n) of said "bifilar" wrapped coils (Figure 7-3) as to (990) of Figure (10-3) encourages the compounding effect (increasing magnetic field-strength during each pulsing cycle) of electromagnetic field-strength (Rp1a xxx Rp1n ~ Rp2a xxx Rp2n) (mutual induction) when applied Pulse-Voltage frequency (49a xxx 49n) of Figure (3-34) passes through the positive energized Resonant Charging Choke (56). Furthermore, the paired coil-wires opposite voltage potential [positive electrical attraction force (B+) ~ negative electrical attraction force (B-)] [hereinafter called Electrical Stress (SS' ~ RR') as to (160) of Figure (3-26)] are always equal in electrical magnitude/intensity since the wire-length of each coil are the same. Pulse-Voltage repetition rate sets up the step-up charging effect Figure (1-3) since the "Resonant Cavity" (Cp) functions as a "Capacitor" (ER) due to the dielectric value of the liquid (or gases) which becomes an integral part of the VIC Circuit, as so illustrated in (650) of Figure (7-4). The

Stanley A. Meyer

RE: Optical Thermal Lens   Memo WFC 429

resultant voltage enhancement (Voltage Amplitude) can exceed 40 kilovolts to instantly convert water (droplets) into thermal explosive energy (gtnt) on demand, as so illustrated in Voltage Intensifier Circuit Diagram (970) of Figure (10-1). Blocking Diode (52) of Figure (4-9) as to Figure (1-1) allows unipolar pulse-wave to go more positive on each pulse-cycle since the Blocking Diode (52) prevents the Resonant Cavity (Cp) from discharging during pulse off-time, as so illustrated in Figure (1-4) as to (60) of Figure (3-22) ... allowing the developed "Electrical Stress"(RU/RU' ~ ST/ST') of Figure (5-1/5-2) across Capacitor Gap (Cp) to go to the farthest point beyond the "State of Equilibrium"... see Atomic Energy Balance of Water (WFC memo 424), once again. The programmable pulse-frequency (49a xxx 49n) of Figure (10-1) input is simply adjusted to tune-in to the dielectric property of the Water Molecule. The resultant Dynamic Electrical Charging Effect (612) of Figure (8-1) acts as a progressive energy enhancer (Energy Priming Stage) (500) of Figure (5-1) when Static State Space (790) of (8-3) is configured to Dynamic State Space (800) of Figure (8-4), as so illustrated in WFC memo (427) titled "Voltage Wave-Guide" ... whereby, the Voltage Wave-Guides forms Water Gap (Cp).

Voltage to Amp Differential Ratio

Opposite polarity Voltage Wave burst (1010) of Figure (10-5) as to Dynamic Voltage Stimulation (770B) of Figure (8-1) is simply produced when Programmable Variable Pulse-Width Pulse-Train Waveform (49a xxx 49n) is allowed to be electrically transmitted through and beyond Resonant Charging Chokes Stages (56/62a xx 56/62n + SS56/62a xxx SS56/62n) of Figure (10-1) that are not only electrically connected in sequential order but likewise magnetically linked by inductance Coupling field (511/512a xxx 511/512n), as so pictorially illustrated in (580) of Figure 6-1). The resultant ever increasing pulsating opposite electrical voltage fields (603/604a xxx 603/604n) of Figure (8-1) having superimposed thereon counter opposing Rippling Voltage-Surfaces (64/B+a xxx 64/B-n) [Dynamic Electrical Charging Effect (612) of Figure (8-1B)], now, set ups, causes, and applies ever increasing (rubberbanding effect) Pulsating Opposite Electrical Stress (RU-RU' ~ ST-ST') across Water Gap (Cp) ... encouraging "Particle Oscillation" as a "Energy Generator" by way of pulsating "Electrical Stress" as the combustible gas atom particles of the water molecule undergo "Particle Deflection" farthest from the point of "State of Equilibrium" and returning back to "Stable State of Equilibrium" during pulse off-time (T_2) for repeated "Snapping Action" (Rubberbanding effect) in accordance with bi-polar Voltage Rippling Effect (1010) of Figure (10-5), as so illustrated in (280) of Figure (3-35). The greater the Electrical Stress (RU-RU'a xxx ST-ST'n) applied (64B+/64B-a xxx 64B+/64B-), the greater amount of thermal explosive energy (16/gtnta xxx 16/gtntn) of Figure (6-2) as to (70) of Figure

RE: Optical Thermal Lens   Memo WFC 429

(4-5) is released from Resonant Water Gap (Cp) (970) of Figure (10-1), as further illustrated in (70) of Figure (4-5). Increasing energy-yield (16/gtnt) still further (xxx 16/gtnt$_{n1}$+ 16/gtnt$_{n2}$ + 16/gtnt$_n$ ... etc.) is accomplished by increasing the number of Resonant Charging Choke Stages (xxx 56/62n + 56/62n1+ 56/62n2 + 56/62n ... etc. ~S~ xxx SS56/62n + SS56/62n1 + SS56/62n2 + SS56/62n ... etc.) of Figure (10-4) in "Sequential Order" (~S~) since the total number of Multi-Coil Magnet bifilar coils (56/62a xxx 56/62n) serially electrically connected together are sequentially electrically linked to an equal number of serially electrically aligned Stainless Steel Resonant Coils (SS/56/62a xxx SS/56/62n) ... allowing each/both bifilar coil assembly (56/62a xxx 56/62n ~S~ SS56/62a xxx SS56/62n) to be electrically and magnetically energized in the same progressive direction toward Water Gap (Cp) and away from blocking diode (55) of Figure (3-34) as to Figure (10-1) and Figure (10-3) ... keeping amp-surge (inhibiting amp flow) to a minimal level [See Voltage Performance Graph (750) of Figure (7-14)] while enhancing Voltage Potential of Electrical Stress (64/RU-RU'a xxx 64/ST-ST'n) as additional Dual Choke Coils (56/62 - SS56/62) are included in the stacked coil-array forming Voltage Intensifier Circuit (970) of Figure (10-1) as to (620) of figure (7-1) ... see Dynamic Voltage Waveform (770) of Figure (8-1), once again.

In Retrospect, the use of Stainless Steel composite coil-wire (430F/FR) consists of both inductance and resistive properties (typically .0048 ohms per foot) which when combined together in metallurgical form aids amp restriction beyond the singularly use of self-inductance magnet (Copper) coil-wire having a lower resistive value. Stainless Steel bifilar Coil-Stage Assembly (SS56/62a xxx SS56/62n) is electrically placed between Magnet Coil-Stage Assembly (56/62a xxx 56/62n) and Water Gap (Cp) to obtain optimum Voltage to Amp Differential Ratio ($V_{highest}$-A_{lowest} ratio). Together, Coil Stages (56/62a xxx 56/62n + SS56/62a xxx SS56/62) added/stacked sequentially into a single overall coil-array assembly (990A/B) of Figure (10-3) forms Amp Inhibiting Network (Figure 8XA) as to (970) of Figure (10-1) (hereinafter called VIC Multi-Coil Spool Assembly).

The magnet Coil-Wire (56/62) is best suited for Voltage inducement while the inductance/capacitance/resistance properties of Stainless Steel coil-wire (SS56~SS62) is appropriately used to restrict electron movement beyond the self-inductance of each energized coil when elevated voltage levels (up to beyond 40 kilovolts) are to be reached/obtained without experiencing any appreciable amount of "Amp Influxing." Generally, magnet coil-wire length is longer than the Stainless steel coil-wire length and magnet bifilar-coil (56/62) is placed on top of Stainless Steel bifilar-coil (SS56/62) to maximize mutual inductance coil-field (Rp2) (adding Rp1

Stanley A. Meyer

RE: Optical Thermal Lens   Memo WFC 429

+Rp2) of (690) of Figure (7-8) to cause coil capacitance (Cda xxx Cdn) to help maintain and even increase pulse voltage amplitude (xxx $V_n + V_{n1} + V_{n2} + V_{n...}$ etc.) while the resistive value (R_{s2}) of SS Coil-Wire (SS56/62) performs the work of further resisting the flow of amps not inhibited by both self-Inductance fields (Rp1 + Rp2), as so illustrated in (690) of Figure (7-8). In all cases, bifilar coils (56/62 ~ SS56/62) are electromagnetically orientated in the same direction.

In terms of operability, electrically flexing (Particle Oscillation) the combustible atoms of the water molecule as a "Energy Generator" by way of opposite voltage polarity is extremely economical since voltage is not consumed in an electronic circuit. Amp Inhibiting Circuit (970) of Figure (10-1) restricts/inhibits amp flow to a minimal level while elevating "Difference of Potential" to the highest possible level. The greater the "Difference of Potential"(in this case, electrical stress) the greater amount of work is performed ... thereby, being in compliance with the Laws of Physics since atoms are the source of all energy in our physical universe and atoms are directly responsive to / stimulated by external electrical forces.

Optical Thermal Lens

When unexcited matter (quiescent state) is stimulated into an excited state (active state) by way of "Electrical Stress", a new form of thermal-energy (Ers) release is possible beyond gas combustion stage (gtnt) by the use of WFC Optical Thermal Lens, as so illustrated in (980) of Figure (10-2). Water Cap (Cp) is, now, Transformed into Vacuum Cavity (Vc) where only hydrogen atoms in a fluid medium is exposed to Voltage Zones (ER) ... forming Capacitor Gap (Cv) since the gas medium, also, exhibits a dielectric value and amp in-fluxing (prohibiting amp leakage into and away from Capacitor Vacuum-Gap Cv) is held to a minimum due to the bifilar wrapped resonant-coils (56/62). The Energy Pumping Stage (520) of Figure (5-3) as to (500) of Figure (5-1) attenuates the hydrogen energy aperture (7) of (550) of Figure (5-8) when applied Opposite Voltage Potential (Va xxx Vn) (49a xxx 49n) (B+/B-) produces pulsating Electrical Stress Wave-form (opposite electrical attraction force) (RU/RU'~ ST/ST'a xxx RU/RU' ~ ST/ST'n) during Pulsing Operations. The stimulated Emitted Energy Radiation (Ers) from the hydrogen atoms are, now, amplified (compressed together) when discrete Emitted Energy Radiation Wavelets (Ersa xxx Ersn) from each hydrogen atom is/are reflected back and forth between Reflective End Plates (915/916) before exiting through Partially Transparent End Plate (916) ... emitting Coherent Energy Wave-Form (917), as so illustrated in (980) of Figure (10-2). The Vacuum Chamber (Cv) containing the hydrogen gas atoms is composed of a high temperature quartz material (918) while Voltage Zones (ER) becomes Voltage Wave-Guides (770) of Figure (8-1) by the use of chemically inert T304 stainless steel material ... forming Electrical Conductance

RE: Optical Thermal Lens   Memo WFC 429

Zones (Skin Effect) (587) ... allowing unipolar Voltage Pulse-waves (583/602a xxx 583/602n) of Figure (8-1A) to travel the linear length of the Voltage Wave-Guide. The Static Electrical Charging Effect (~) (585) is set up when applied Electrical Stress Wave-form (RU/RU' ~ ST/ST') is simultaneously applied on opposite sides of the hydrogen atoms ... propagating "Particle Oscillation" as a "Energy Generator" in a vacuum rather than by chemical interaction (chemical stress) of gas combustion (gtnt).

Quartz Tube Configuration & Operational Parameters

Exposing the hydrogen atoms to applied Static Voltage Stimulation (770) of Figure (8-1) causes the Static Electrical Charging Effect (585) to set up "Voltage Tickling of State Space" which takes the hydrogen atom (s) from "Quiescent State"(Qs) to "Active State"(As) and then back again to "Quiescent State (Qs) once applied unipolar pulse-wave (583/602) goes through Voltage Pulsing Cycle (Vpwf) from "Ground State"(Gs) to Voltage Peak potential (Vpp) (780) of Figure (8-2A) and then returns to "Ground State" (Gs) for continued repetitive Voltage-Pulsing (583/602a xxx 583/602n) ... forming Pulse Wave Frequency (Pwf), as so illustrated in (780) of Figure (8-2) as to (770) of Figure (8-1).

The resultant Flex-Density (Fd) of the hydrogen atom (s) is, now, directly related to applied Voltage Amplitude (64a ~ 64b ~ 64c ~ 64n) of Figure (8-2A) which directly determines the Field Strength (Fs) of Static Electrical Charging Effect (Sece)(585). The Voltage Pulse Field (Vpf) (forming leading and trailing edges of each Voltage Pulse) determines the duration of the Static Electrical Charging Effect (Sece) being superimposed onto the Hydrogen Atom (s) during each Voltage Pulsing Cycle (Vpwf). The Electrical Conduction Zone (587) (Skin Effect) between the dielectric gas medium (919) and the Electrical Contact Surface of the inside surface area of the Voltage Wave-Guides (66/67) allows Unipolar Pulse Train (583a/602a ~ 583b/602b ~ 583n/602n) to travel the entire length of the Voltage Wave Guides that make up Voltage Zones (66/E9 ~ 67/E10). Static Voltage Stimulation (770) of Figure (8-1A) is where Voltage Peak Potential (Vpp) remains constant during Voltage Pulse Formation (Vpb/Vpa) to keep reforming Flex-Density Potential (Fdpa xxx Fdpn) from going beyond a certain point away from Static State of Equilibrium (Esse) [Quiescent State of the hydrogen atom (s)]; whereby, Dynamic Voltage Stimulation (770) of Figure (8-1B) continues to go farther and farther away from the State of Equilibrium (Esse) during each and every Voltage Pulsing Cycle (Vpf), as so illustrated in (770) of Figure (8-1B) ... establishing variable Dynamic Electrical Charging Effect (612).

Stanley A. Meyer

RE: Optical Thermal Lens   Memo WFC 429

In both cases, Static Voltage Stimulation (770) of Figure (8-1A) and Dynamic Voltage Stimulation (770) of Figure (8-1B) incorporates the use of Positive Electrical Voltage Potential (B+) (E11) and Negative Electrical Voltage Potential (B-) (E12) to form synchronized diametrically opposed Voltage Gate-Pulse (Vgp) (583/602) across Vacuum Gap (Vc) ... thereby, establishing functional parameters of Optical Thermal Lens (980) of Figure (10-2) when Voltage Intensifier Circuit (VIC Circuit) (10-1) is electrically connected to Voltage Zones (ER) (66/E11~67/E12). Remember, Voltage-Sync Gate-Pulse (Vgp) produces opposite electrical attraction forces (B+/ST-ST' ~ B-/RU-RU') of Figure (5-1) that are not consumed in an electronic circuit.

A Technique Called "Easer"

Energy Priming Stage (500) of Figure (5-1) is, now, activated and performed when the hydrogen atom (s) is exposed to applied Voltage-Sync Gate Pulse (Vgp) by which Electrical Voltage Attraction Forces (Electrical Stress) (RU-RU' ~ ST-ST') causes "Particle Oscillation" of the hydrogen atom (s) to emit radiant energy (Ers) (919), as so illustrated in (980) of Figure (10-2) as to (500) of Figure (5-1). The applied unipolar Positive Voltage Pulse (B+) (ST-ST') electrically attracts the negative charged atom electron; while, simultaneously, applied unipolar Negative Voltage Pulse (B-) (RU-RU') electrically attracts the positive charged Proton (3) that forms the nucleus of the hydrogen atom ... causing the hydrogen atom to elongate under Atomic Electrical Stress (Aes)($\Delta AA' \sim \Delta ZZ'$)... whereby, atomic electrical attraction force (4) that exists between the deflected orbital negative charged electron (1) and the pivotal positive charged Proton (3) is attenuated ($\Delta 4a$ xxx $\Delta 4n \sim \Delta 5a$ xxx $\Delta 5n$) ... which, in turns, attenuates the spin-velocity of the "Gyroscopic Regulator" of the hydrogen nucleus to cause Energy Aperture (7) to emit/transmit more or greater amount of Universal Energy (9) into, through, and beyond the energy spectrum of the hydrogen atom for "Energy Propagation" (atoms being an "Energy Generator" in our physical universe) by way of "Particle Oscillation" of the hydrogen atom (or any other atom), as so illustrated in (570) of Figure (5-10) as to (550) of Figure (5-8).

Voltage Peak Potential [peaking-out electrical stress (xxx Aes) since opposite voltage potential creates/determines electrical field-strength/establishes electrical attraction forces between electrical charged particles in space relationship in an electronic circuit] (Vpp) not only determines on how far the deflected orbital electron will move away from the pivotal proton during "active state" but, also, is directly related to Flex-Density (Fd) of the hydrogen atom ... Flex-Density (Fd) being the measure of energy-intensity (Wavelets of energy) emitted/released from the oscillatory

RE: Optical Thermal Lens
Memo WFC 429

hydrogen atom (undergoing particle oscillation) in proportional relationship to the field-strength/intensity of the external electrical stress-force (Esf) superimposed onto the electrical field strength (AA'~ZZ') of Figure (5-3) of the hydrogen atom in "Quiescent State-Space" (Qss) ... which, in turns, causes Energy Aperture (7) of the hydrogen atom to elongate (enlarges) to emit a discrete amount of Universal Energy (Ue)(9) into the energy spectrum of the hydrogen atom during each and every applied Voltage-Sync Pulsing Cycle (Vgpa xxx Vgpn).

Going from "Quiescent State-Space" (Qss) to "Active State-Space" (Ass) and returning to "Quiescent State-Space"(Qss) once again for repetitive "State Change" beyond the "State Condition" of Equilibrium (Esse) in a given/predetermined unit of time (Δ time) is, herein, called "Voltage Tickling of State Space." The greater the electrical stress force (xxxEsfn) per (~) Voltage Gate-Pulse (Esf~Vgpa xxx Esf~Vgpn) applied, the greater amount of Flex-Density (xxx Fdn) of electromagnetic radiant energy (917a xxx 917) of Figure (10-2) per cm^3 occurs.

Once the hydrogen atom "absorbs" a sufficient amount of Universal Energy (Ue)(9a~9b~9c~9n) within the atom energy spectrum, the deflected electron (15) that have moved to a higher energy state (activated state)(K to L orbit or beyond) suddenly jumps back to it original lower energy state (quiescent state)(back to K orbit) when Input Pulse Frequency (49) terminates Voltage-Sync Pulse (Vsp) during applied pulse off-time (T2) of Figure (7-8) ... causing spontaneous emission of coherent energy (919) of Figure (10-2) to be emitted from partially transparent end plate (916) as long as positive pulse "on-time" (T1) of Figure (7-8) continues. The Radiant-Intensity (Rei) of the coherent wave-energy (919) being released from the quartz tube (918) is, further, enhanced when emitted energy-wavelets (Ers) given off by the activated hydrogen atom are allowed to oscillates (back and forth movement) at an ever increasing "difference of potential" between the end plates (915/916).

Together but in separate variable forms, Applied Voltage Peak Potential (Vpp), Voltage Pulse Cycling (49a xxx 49n), and Opposite Voltage Potential (Vgpa xxx Vgp) forming Voltage Sync-Pulse (49~Vpp~Vgpa xxx 49~Vpp~Vgpn) determines the energy-intensity (Eie) of the coherent-beam (919). Focusing lens (921) is simply used to redirect the Radiant-Energy (917) to a heat diffuser (923) capable of converting Radiant-Energy (917) to heat energy for industrial usage. The stimulated spontaneously emission of electromagnetic radiation from the hydrogen atom (or other atoms) by way of "Electrical Stress" is, hereinafter, called "Easer."

Stanley A. Meyer

RE: Optical Thermal Lens　　　　　　　　　　　　　　　　　　　　　Memo WFC 429

970

Electrical Stress
(Opposite Electrical Attraction Force)
Water Dielectric Value 78.54Ω

Positive Electrical Voltage zone (B+) (66/E9)

Negative Electrical Voltage zone (B-) (67/E10)

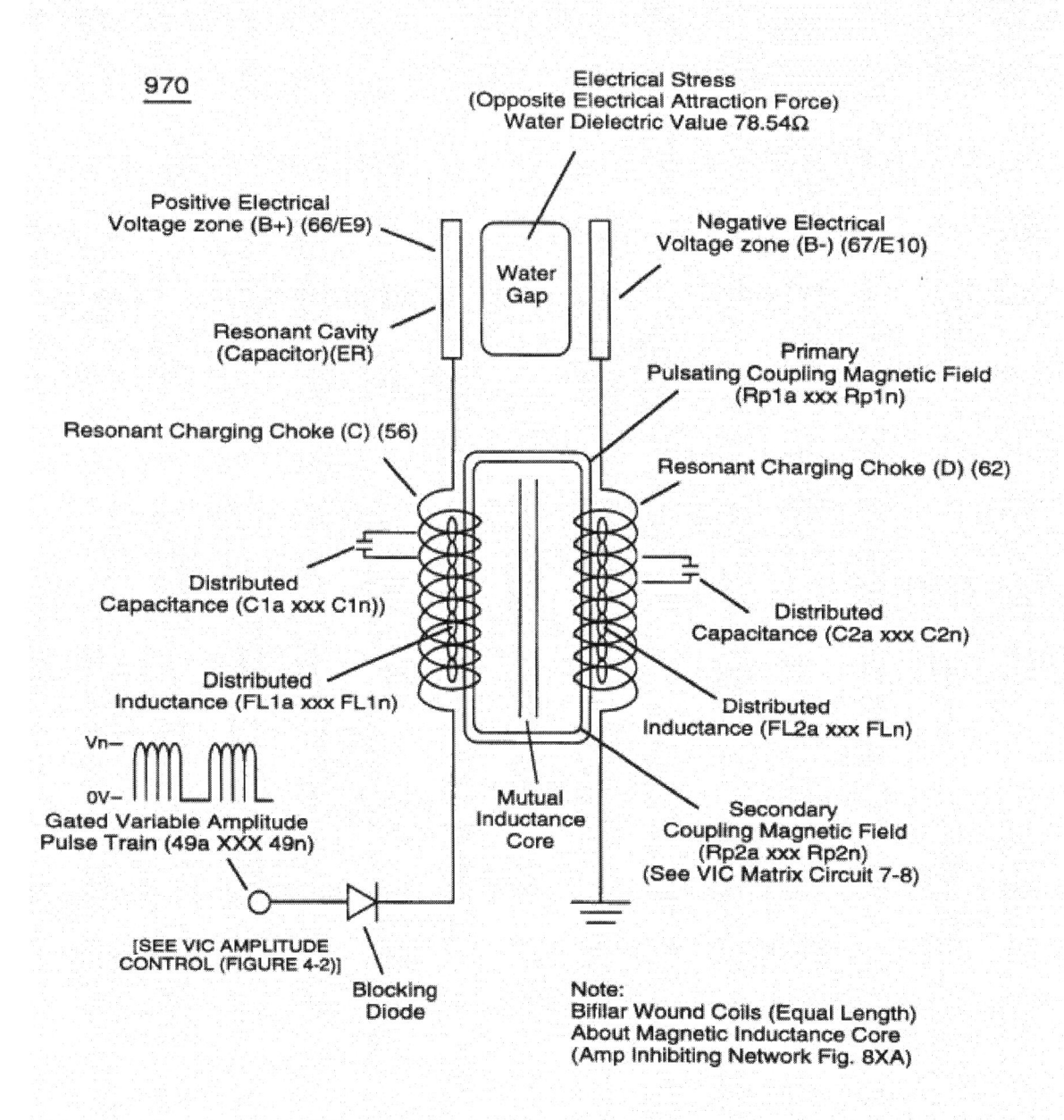

Water Gap

Resonant Cavity (Capacitor)(ER)

Primary Pulsating Coupling Magnetic Field (Rp1a xxx Rp1n)

Resonant Charging Choke (C) (56)

Resonant Charging Choke (D) (62)

Distributed Capacitance (C1a xxx C1n)

Distributed Capacitance (C2a xxx C2n)

Distributed Inductance (FL1a xxx FL1n)

Distributed Inductance (FL2a xxx FLn)

Vn−
0V−
Gated Variable Amplitude Pulse Train (49a XXX 49n)

Mutual Inductance Core

Secondary Coupling Magnetic Field (Rp2a xxx Rp2n)
(See VIC Matrix Circuit 7-8)

[SEE VIC AMPLITUDE CONTROL (FIGURE 4-2)]

Blocking Diode

Note:
Bifilar Wound Coils (Equal Length)
About Magnetic Inductance Core
(Amp Inhibiting Network Fig. 8XA)

FIGURE 10-1: VOLTAGE INTENSIFIER CIRCUIT

Stanley A. Meyer

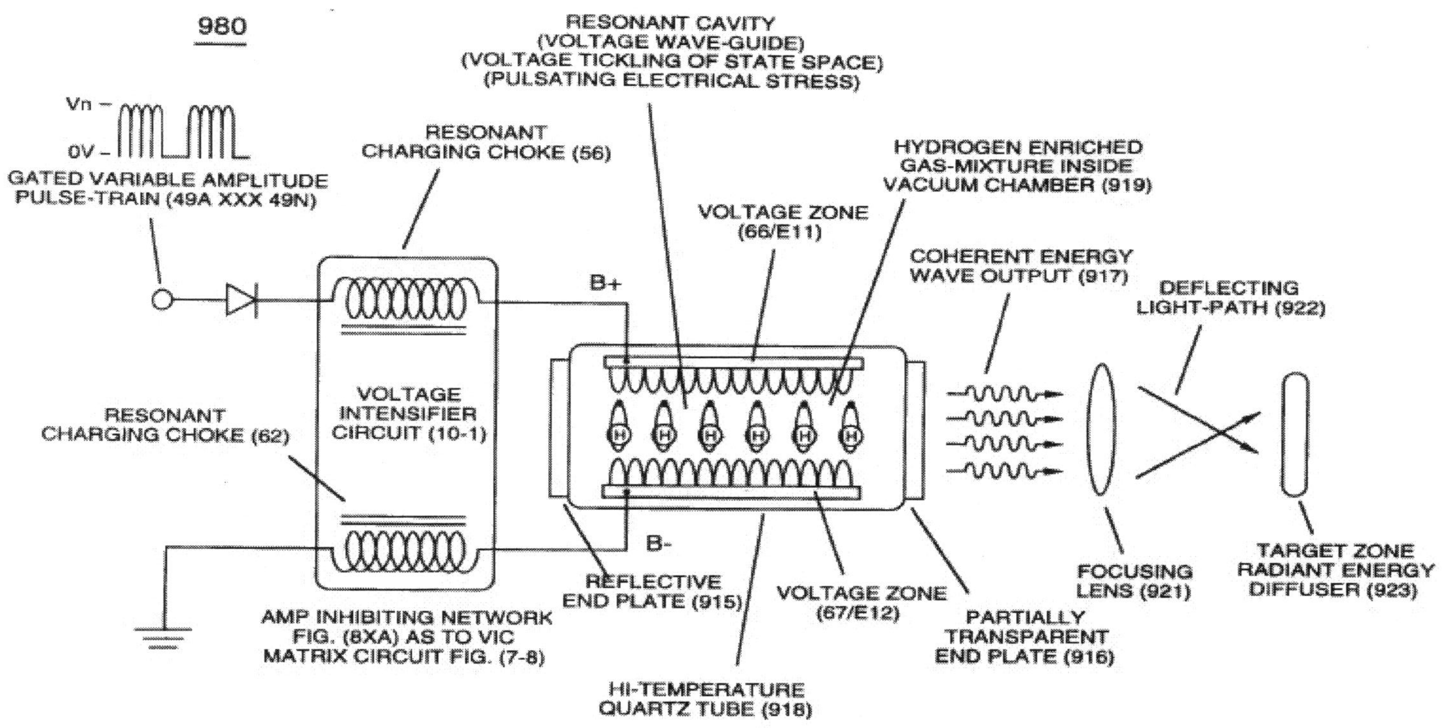

FIGURE 10-2: OPTICAL THERMAL LENS

RE: Optical Thermal Lens   MEMO WFC 429

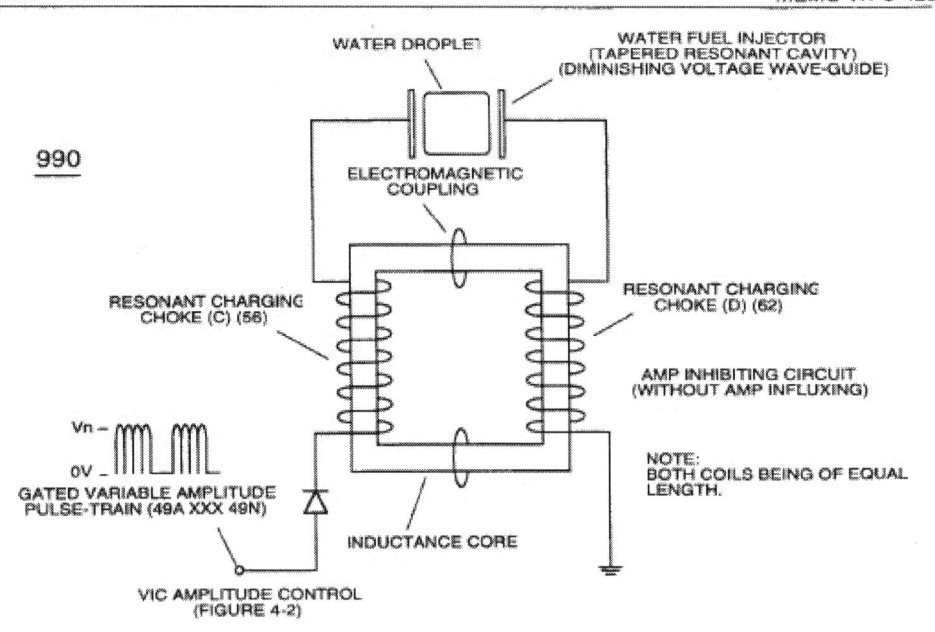

990

FIGURE 10-3A: VIC DUAL SINGLE-COIL ASS'Y

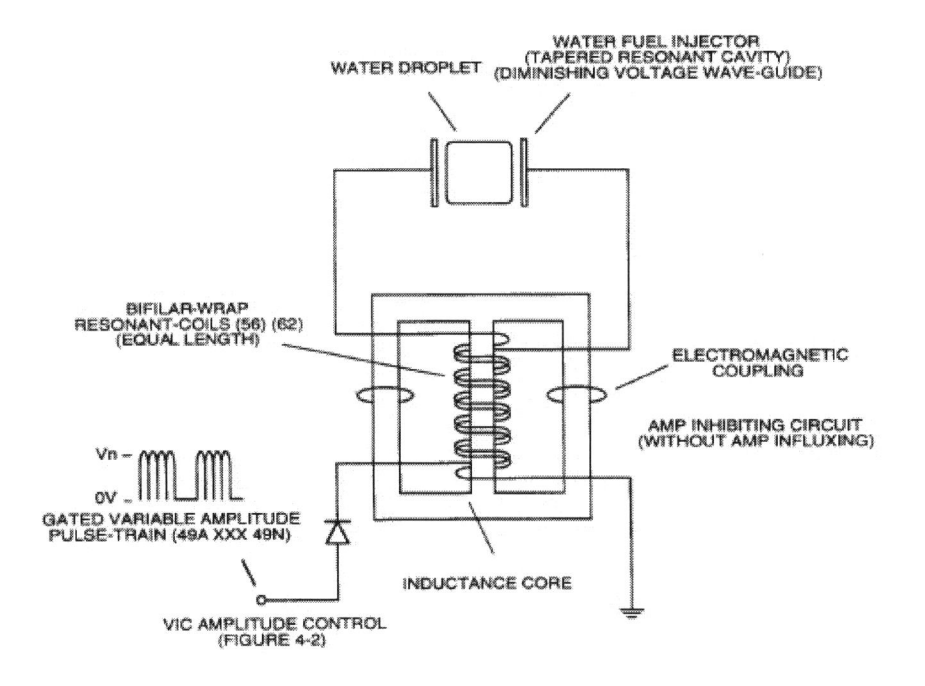

FIGURE 10-3B: VIC BIFILAR-WRAP ASS'Y

FIGURE 10-3: VIC COIL-WRAP CONFIGURATION

Stanley A. Meyer   10-11

RE: Optical Thermal Lens   Memo WFC 429

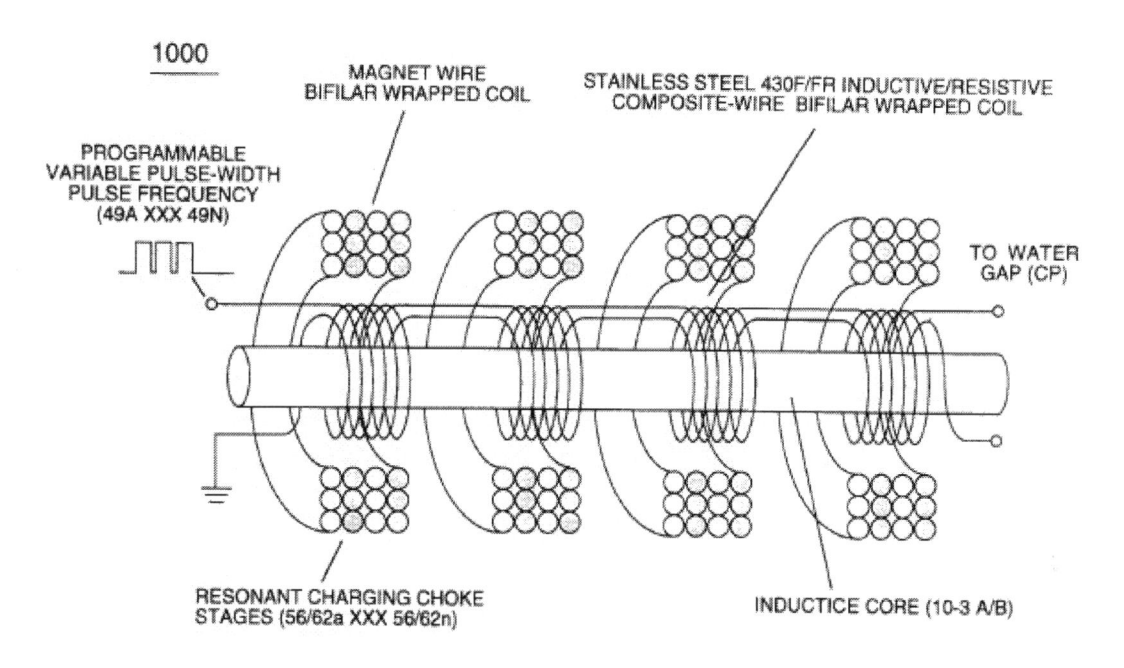

1000

- PROGRAMMABLE VARIABLE PULSE-WIDTH PULSE FREQUENCY (49A XXX 49N)
- MAGNET WIRE BIFILAR WRAPPED COIL
- STAINLESS STEEL 430F/FR INDUCTIVE/RESISTIVE COMPOSITE-WIRE BIFILAR WRAPPED COIL
- TO WATER GAP (CP)
- RESONANT CHARGING CHOKE STAGES (56/62a XXX 56/62n)
- INDUCTICE CORE (10-3 A/B)

FIGURE 10-4: DUAL-LAYERED MULTI - SPOOL CONFIGURATION

1010

- VOLTAGE WAVEBURST STEP CHARGING EFFECT (PARTICLE OSCILLATION AS A ENERGY GENERATOR)
- RIPPLING POSITIVE VOLTAGE POTENTIAL
- B+
- DIFFERENCE OF POTENTIAL (EVER EXPANDING PULSATING ELECTRICAL STRESS) ACROSS WATER GAP (CP)
- RIPPLING NEGATIVE VOLTAGE POTENTIAL
- B-

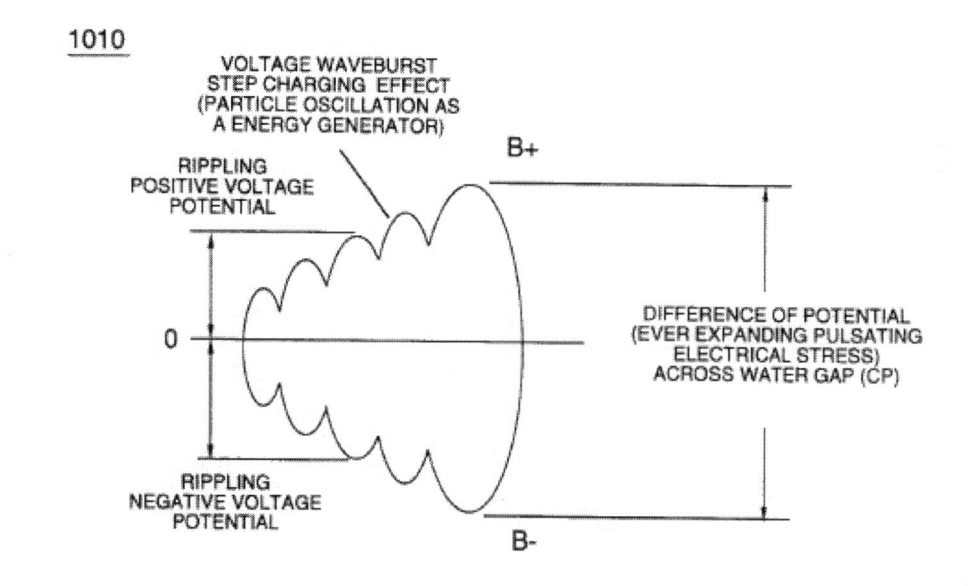

FIGURE 10-5: OPPOSITE VOLTAGE CHARGING EFFECT

Stanley A. Meyer   10-12

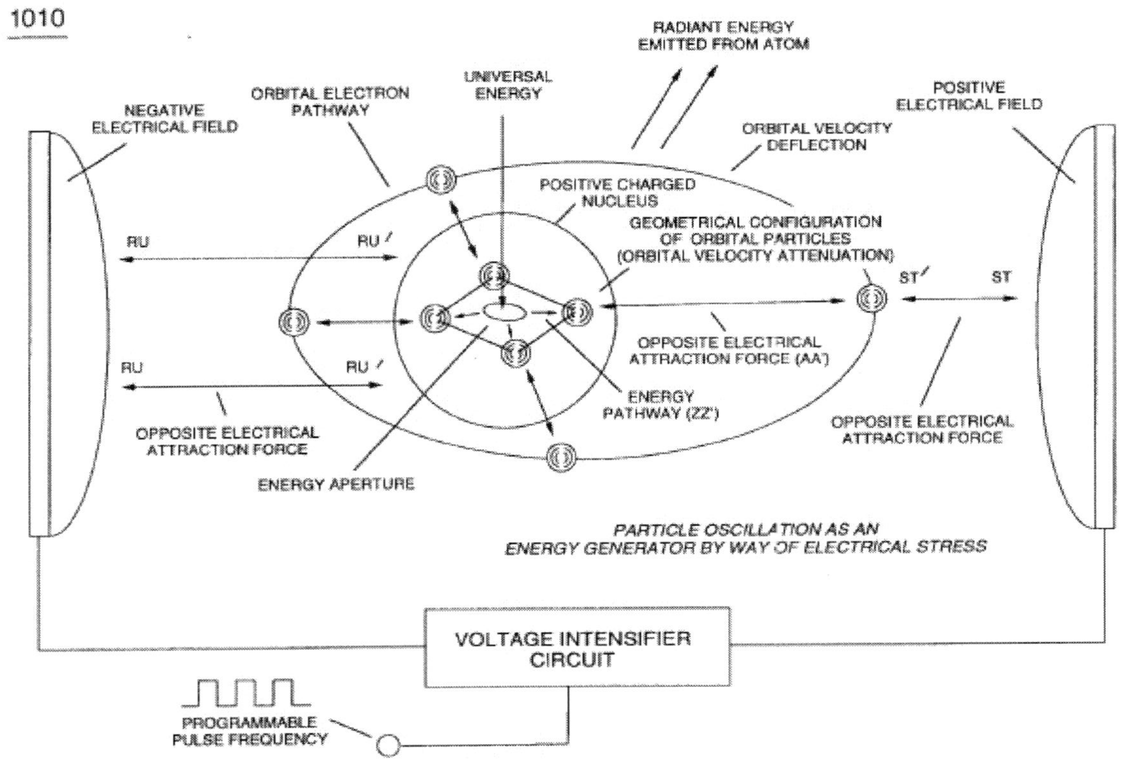

FIGURE 10-6: VOLTAGE TICKLING OF STATE SPACE

Memo WFC 430

WFC Steam Resonator

Particle Oscillation as a Energy Generator

Flexing the atom as a energy generator is as basic as science itself ... for without the atom having the capability to release energy our universe would not exists as we know it.

Atoms can be "Flexed" in many different ways to release a certain/given type of energy. Physical Stress being subjected to and undergoing Electrical Stress is by far the most dynamic way to cause the atom to go farthest from the point of state of equilibrium without atomic decay.

Electrical Stress brought on by opposite voltage polarity in a electronic circuit is extremely economical since applied voltage fields propagating opposite electrical attraction forces between the bipolar water molecule and the stationary voltage fields are not consumed in an electronic circuit when amp influxing is prohibited.

The Voltage Flexing Process to deflect the water molecule under both physical and electrical stress to emit thermal heat energy from the atom (s) of the water molecule under control state, is, now, to be presented by the utilization of the WFC Steam Resonator technology which incorporates the use of the Voltage Intensifier (VIC) Switchover Circuit to cause "Particle Oscillation" as a "Energy Generator:"

RE: Steam Resonator   Memo WFC 430

Steam Resonator

Particle Oscillation As An Energy Generator

All energy in our physical universe (The third dimension) comes from a singular source ... the atom. There are four basic forces that make up and effect the atomic structure: electrical force, electromagnetic force, weak and strong nuclear forces, and gravity. By either attenuating either one or more of these atomic forces, energy can be release from the atom to perform work in a variety of ways: such as, emitting photon, electromagnetic, or even radiant heat energy. Exposing the water molecule atom (s) to an external electrical attraction force (SS'/RR') separately or combining the external electrical attraction force with an external electrical repelling force (SS'~TT'/RR'~WW') can cause the bipolar electrical charged water molecule atom (s) to release thermal heat energy when physical impact (physical force) is achieved as a result of particle (s) colliding together under electrical stress which becomes and is the physical mover ... causing electron bounce to oscillate the energy aperture of each atom of the water molecule.

Voltage Flexing Process

Particle oscillation as a "Energy Generator" by way of "physical impact" caused by a singular unipolar voltage pulse wave-form alternately polarity triggered is yet another method beyond the prior art to flex the water molecule to release thermal energy (Kinetic Energy) from the water molecule atom (s) without the need of gas combustion brought about by gas separation from water, as so illustrated in (1050) of Figure (11-5).

In order to accomplish this task, dual unipolar voltage pulse circuit (1010) of Figure (11-1) is, now, utilized to deflect (Physical Movement) the bipolar electrically charged water molecule (210) of Figure (3-46) while undergoing and experiencing both physical and electrical stress, simultaneously ... causing atomic flexing of the water molecule atom (s) energy aperture (7) of Figure (5-1) which, in turns, releases radiant thermal heat energy (165) from the atom structure (s), as further illustrated in (450) of Figure (3-46).

As applied external opposite electrical attraction forces (S-S') and/or (R-R') as so shown in (1030) of figure (11-3) captures and electrically locks onto either the negative charged oxygen atom or onto the positive charged hydrogen atom (s) ... whichever the case may be, the applied stationary voltage fields (952/E13 ~ 953/E14) or (954/E15 ~ 956/E16) alternately switch over periodically superimposes electrical stress forces (S-S' and R-R') onto the energy spectrum of the water molecule atom (s)(210) while physical flexing (951) of Figure (11-5) of the water molecule atom (s) occurs ... disrupting the spin-velocity of

Stanley A. Meyer   11-1

RE: Steam Resonator   Memo WFC 430

water molecule atom (s) orbiting electrons (s) ... forcing energy Apertures (7) of both unlike atoms of the water molecule (500) of Figure (5-1) and (510) of Figure (5-2) to be momentarily enlarged to a greater size (Particle flexing ... called hereinafter Particle Oscillation), separately but simultaneously ... allowing a greater amount of energy to enter into, travel through, and pass beyond the energy spectrum of each stimulated atom (s), respectfully ... emitting the additive/surplus energy away from the excited atom (s) in the form of radiant thermal heat energy (165) when the flexed atom (s) (undergoing physical/electrical stress) returns to stable state of atomic equilibrium once applied electrical pulse-voltage wave-form (952 ~ 953) or (954 ~ 956) is electrically switch off and permitted to collapse back toward electrical ground state of zero volts (0V).

Repetitive formation of pulse voltage fields (952a xxx 952n) ~ 953a xxx 953n) or (954a xxx 954n~ 956a xxx 956n) continues this "Voltage Energized Thermal Transference Effect" (1050) of Figure (11-5) (hereinafter called Atomic Flexing Process) during each and every pulse voltage on-time, as so illustrated by way of gated pulse-voltage waveform (1020) of Figure (11-2). In essence, then, the continued flexing of a liquid or gas atoms being exposed to physical stress (954) by an external electrical attraction force (S-S'/R-R') is, herein, a more effective way to induce and propagate "Particle Oscillation" as an "Energy Generator" since voltage potential of opposite polarity poses a greater "Differential of Potential" over the prior art. (See Memo WFC 429 titled " Optical Thermal Lens" as to Memo WFC 424 titled "Atomic Energy Balance of Water" for further references).

VIC Switchover Circuit

VIC Switchover Circuit (1010) of Figure (11-1) is utilized to bring about Voltage Flexing Process (1050) by preventing amp influxing into and away from the separate and periodically spaced (~) voltage zones (E13-E14 ~ E15-E16) of diagram (1030) of Figure (11-3) as to (1010) of Figure (11-1); while, at the same time, allowing voltage potential of opposite electrical attraction forces (S-S'/R-R') to perform work by deflecting the bipolar electrical charged water molecule (210) in a given directional pathway, as so in accordance to/with Coulomb's (Eq12) and Newton's second law of electrical force (Eq13) in an electrical/electronic circuit.

When incoming programmable gated pulse-frequency waveform (T4A) of Figure (11-2) electrically energizes primary input coil (957) of Figure (11-3) to produce positive voltage field (952) across voltage zone (E13), the bipolar electrical charged water molecule having a negative charged oxygen atom is deflected and moved toward stationary positive voltage plate (E13) due to the opposite electrical attraction force (S-S') that exists between both opposite electrically charged entities. Whereas, in like manner and in the same instant of time, stationary negative voltage field (958) attracts and displaces another and totally separate bipolar water molecule in an linear movement since opposite electrical attraction force (R-R') also exists between stationary negative charged voltage plate (953) and the, now, moving positive charged hydrogen atom (s) being electrovalently linked to the negative charged atom.

The resultant physical displacement (physical movement) of both separate bipolar water molecule moves relatively at the same displacement velocity since both particle masses of the water molecule (s) are basically identical in volume-size and the electrical intensity on both stationary voltage fields (952/953) are similar due to the fact that both primary coil (957) and secondary coil (958) comprising and forming voltage intensifier circuit (990) of Figure (10-3) are together bifilar wrapped in equal length.

The simultaneously formation of both the positive voltage field (952) and the negative voltage field (953) is simply accomplished by the mutual electromagnetic inductance coupling field that is produced between the two bifilar wrapped coils (957/56 ~ 958/62) when the primary coil (957/56) is electrically energized by incoming voltage pulse train (T4a xxx T4n), as so illustrated in (970) of Figure (10-1).

Automatically, the self-inductance coupling (619a xxx 619n) of Figure (7-3) prevents amp influxing [restricting current flow into and away from water bath (68) during each pulsing cycle T4A/T4B]. While, the distributed capacitance (Cda xxx Cdn) of each coil experiencing inductance coupling (619) elevates applied voltage level (Vn) to a higher voltage amplitude (increasing voltage intensity) required to deflect the bipolar water molecule to a given or preselected distance. Voltage intensity (952/953) is, therefore, directly determined by the number of turns of each coil (957/958) as to the applied voltage amplitude of incoming pulse-wave (... xxx Vn) (1060) of Figure (11-2c). Voltage intensity as in terms of "Different of Potential" establishes the amount of work performed by the applied "Electrical Stress" to bring about molecule mass displacement of the water molecule in a liquid medium.

Electrically energizing Voltage Intensifier (VIC) Circuit (1002), now, causes both bipolar water molecules (1004/1006) to be deflected and displaced in a left-hand movement, as so illustrated in (1030) of Figure (11-3). To reverse direction of the line of travel of the deflecting water molecule from left-hand movement to right-hand movement, another and completely separate Voltage Intensifier VIC-Coil Assembly (1003) of Figure (11-3) is periodically switch-on electrically by alternate voltage pulse waveform (T4B) once first voltage pulse wave-form (T4A) is terminated for a brief period of time (T3A) ... duplicating the electrical attraction force (S-S'/R-R') as before except both bipolar water molecule (s) (1004/1006) are, now, redirected and deflected in the opposite direction toward voltage fields (956/954), respectfully, as so illustrated in (1030) diagram (B) of Figure (11-3).

RE: Steam Resonator Memo WFC 430

This continued and repeated oscillation of the bipolar water molecule (1004/1006) in opposite direction of linear travel (back and forth motion) produces kinetic energy (165) when the moving and deflected bipolar water molecule (1004/1006) or any other bipolar molecule of water interlocking with ever changing electrical attraction forces (S-S'/R-R') collides with neighboring water molecules present in the same water bath (68).

Electrically interfacing alternate "Switchover" voltage pulse wave-form (T4A/T4B) to each of both VIC Coil-Array (s) (1002/1003) of Figure (11-3) as schematically depicted, now, forms VIC Switchover Circuit (1010) of Figure (11-1).

Pairing together positive voltage zone (E13/952) with negative voltage zone (E16/956) and doing the same with voltage-surfaces (E14/953) to (E15/954) as so graphically shown in (1010) of Figure (11-1) and each having an longitudinal axis of identical length, now, individually forms what is called hereinafter a "Differential Voltage Wave-guide"(1040) ... being defined as heating water by alternate pulsation of opposite voltage fields at different pulse time-on periods, collectively called "Voltage Switchover Firing Logic (B+/0 - B-/0 ~ 0/B - 0/B+).

In Like manner as to linear cylindrical resonant cavity (730) of Figure (7-12), the Differential Voltage Wave-Guide (1040) of (1010) of Figure (11-1) as to Figure (11-6) is constructed in such a way as to allow a smaller tube to be placed inside a much larger tube having space relationship to allow water to pass there between, as so pictorially shown in (170) of Figure (3-25). The constructed tubular-Array is composed of T304 stainless steel material, or any equivalent thereof, which is chemically inert to the voltage deflecting process (1050) of Figure (11-5).

The electrical conductivity of the stainless steel material T304 and the dielectric properties of water (water being an insulator to the flow of amps) both together sets up electrical conduction zone (587) which aids the ability of the voltage pulse waveform (T4a xxx T4n) and/or (T4ba xxx T4bn), whether be it positive or negative in electrical polarity, to be electrically transmitted and linearly displaced along the longitudinal axis of the inner side walls adjacent to water bath (68), as so illustrated in (1040) of Figure (11-4). This phenomenon of transferring voltage waveforms along an electrical conductive surface is known in the field of physics as the Skin Effect. The dielectric value of water (78.54) inhibits amp leakage into the water bath ... preventing distortion of the reoccurring and traveling voltage waveforms (T4Aa xxx T4An/ T4Ba xxx T4Bn). The resultant pulsating electrical stress (S-S'/R-R) penetrates the liquid bath of water since the water bath takes on an electrical charge when the rotational spin (1019) of the water molecule (s) occurs to bring about bipolar alignment of the water molecule comprising water bath (68)

RE: Steam Resonator   Memo WFC 430

during each and every reversing voltage pulsing cycle (B+/0 - B-/0 ~ 0/B+ - 0/B-), as so illustrated in (650) of Figure (7-4).

Not only does the alternate first gated voltage pulse (B+/0 - B-/0) and then the second gated voltage pulse (0/B+ - 0/B-) oscillates the bipolar water molecule (s) back and forth in rapid succession to produce heated water at a predetermined temperature level on demand; but, also, deflects the oscillating bipolar water molecule in an upward direction since the reforming voltage pulse waves are always in a state of progressive movement of linear displacement ... performing the same function as a water pump... a water pump, however, not having any mechanical moving parts to wear out.

Varying the gated pulse-width (T4A/T4B), attenuating voltage amplitude (xxx Vn), and the Switchover pulse frequency rate (Sopr), collectivity determines the rate by which the water temperature rises as the water medium (68) travels through the Differential Voltage Wave-Guild (1040), as so illustrated in (1060) of Figure (11-2) as to (1040) of Figure (11-6).

Electrical Crossover Switching Circuit

To reduce the number of Voltage Intensifier VIC-Circuit to the use of only one VIC Coil Assembly (1002) while encouraging the utilization of using a Voltage Repelling Force (W-W') and/or (T-T'), Electrical Crossover Circuit (1060) of Figure (11-7) is, now, electrically placed between VIC-Coil Assembly (1003) and both Differential Voltage Wave-Guilds (1030A/1030B), as so illustrated in (1050) of Figure (11-7). The Electrical Crossover Switching Circuit (1060) singularly places either a Positive Voltage Potential (1014) across both Voltage Zones (E18/E14) and/or a Negative Voltage Potential across Voltage Zones (E17/E16) or vise versa. In doing so, Electrical Repelling Forces (T-T') and (W-W'), now, exerts a "Pushing Effect" onto the already deflecting water molecules (1017/1018) since like electrical forces repel or push away from one another in a strictly physical manner.

In terms of operational parameters, Electrical Attraction Force (S-S'/R-R') and Repelling Forces (T-T'/W-W') can be applied simultaneously or applied in a time sequence of events as Electrical Crossover Switch Circuit (1060) reverses the voltage polarity from one Differential Voltage Wave-Guild (1040B) to another and completely separate Differential Voltage Wave-Guild (1040A) of similar or like configuration ... preforming Voltage Switchover Logic Functions (B+/B+/1030B ~ B-/B-/1030A Switchover B-/B-/1030B ~ B+/B+/1030A) of Figure (11-7) during each and every sequential voltage pulsing cycle (T4A ~ T4B ~ T4A ~ T4B and so on). When (B+/B+ ~ B-/B-/1030B) switch function is activated, switch terminals (T1/T2 ~ T3/T4) are closed. Switch position (T1/T4 ~ T3/T2) reverses voltage

Stanley A. Meyer   11-5

RE: Steam Resonator   Memo WFC 430

polarity once switch function (T1/T2 ~ T3/T4) goes to close position after Switch Logic Function (1013) becomes an open circuit ... and then vice versa and so on in an repetitive format ... causing "Particle Oscillation" as a "Energy Generator" by way of "Physical Stress" undergoing pulsating "Electrical Stress" whenever pulse switching cycles (1060) is electrically activated by incoming trigger pulse frequency (1019/T4a xxx 1019/T4n), as so illustrated in (1050) of Figure (11-7). Oscillating the bipolar water molecule by way of opposite voltage fields without amp influxing to heat water on demand, hereby, defines the "Mode of Operability" of the WFC Steam Resonator. Δ

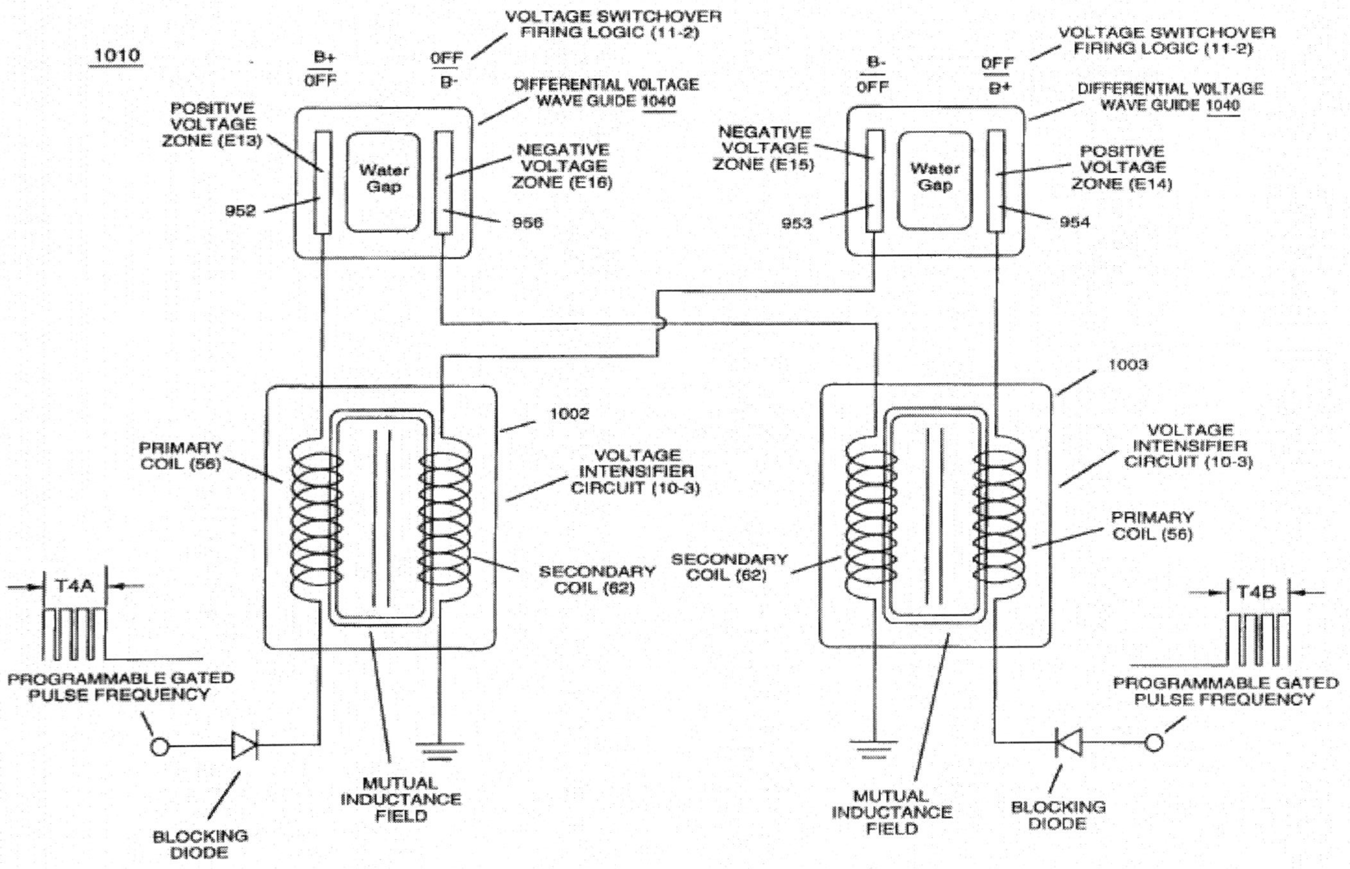

FIGURE 11-1: VIC Switchover Circuit

RE: Steam Resonator Memo WFC 430

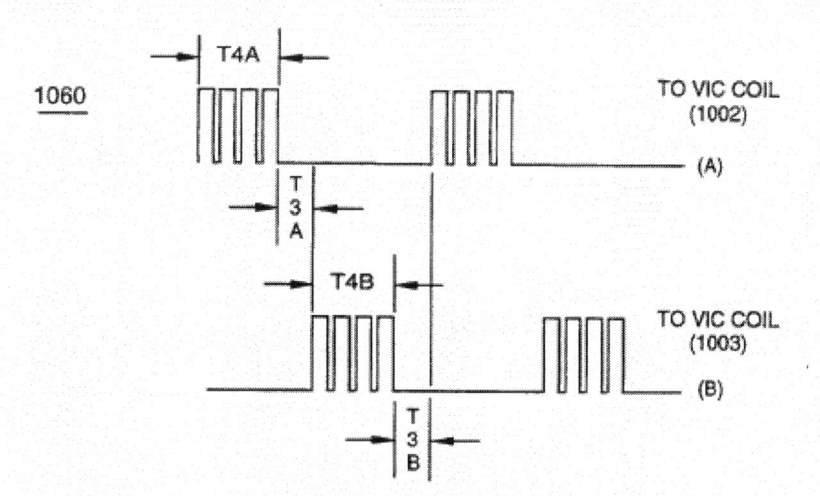

1060

(A) ALTERNATE-FIRING PROGRAMMABLE
GATED PULSE FREQUENCY

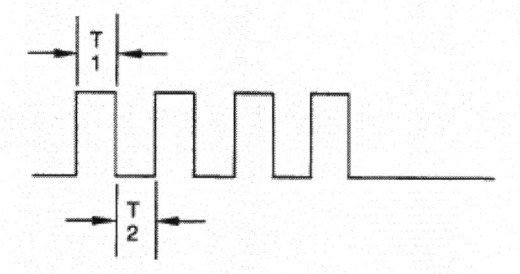

(B) PROGRAMMABLE PULSE FREQUENCY

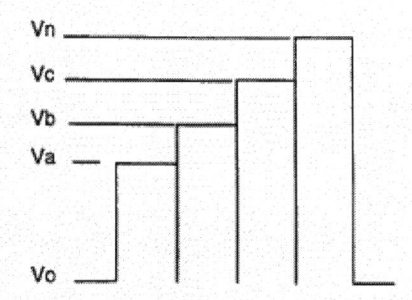

(C) DIGITAL VOLTAGE AMPLITUDE CONTROL

FIGURE 11-2: Voltage Switchover Firing Logic

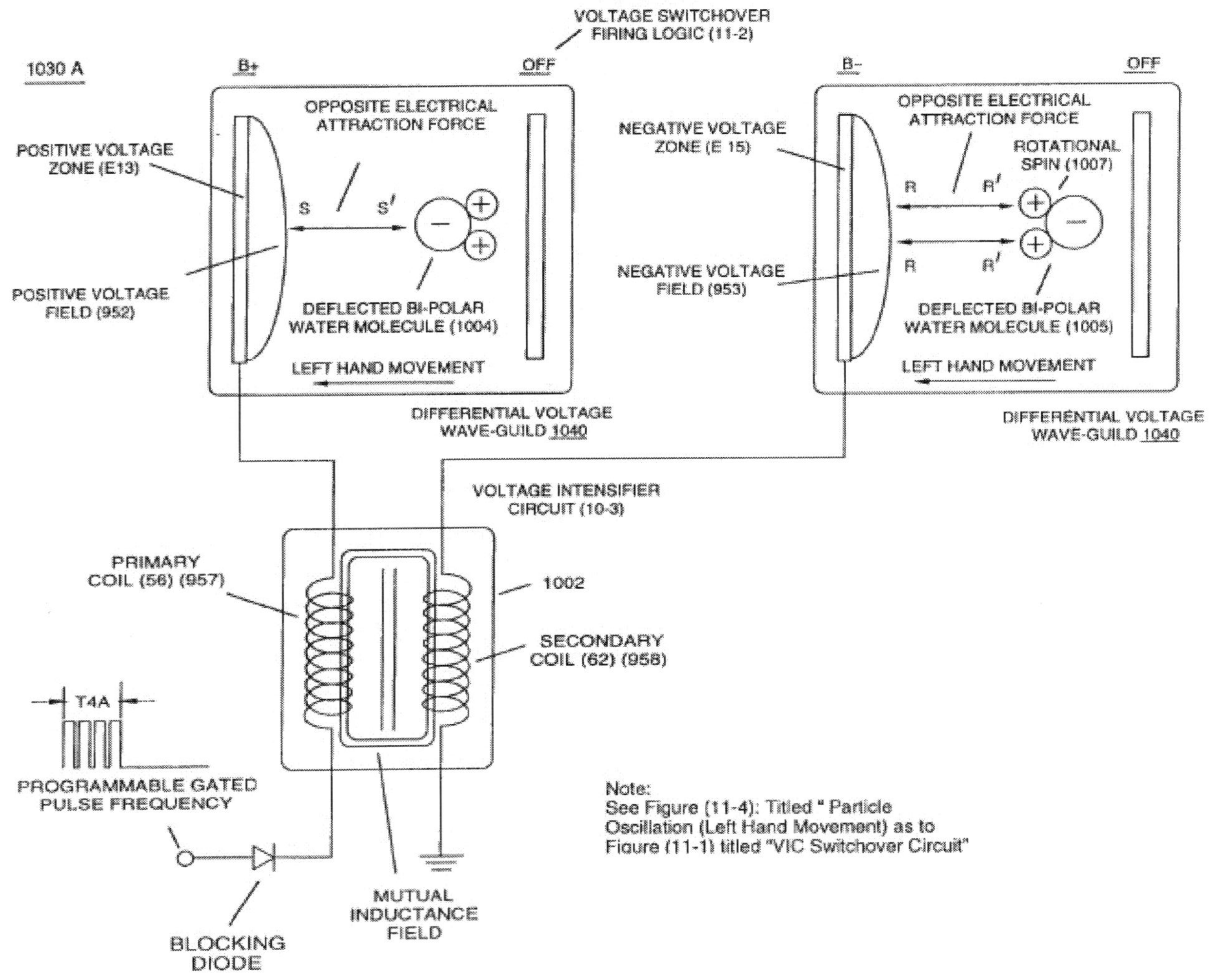

FIGURE 11-3: PARTICLE OSCILLATION (LEFT HAND MOVEMENT)

Note:
See Figure (11-4): Titled " Particle Oscillation (Left Hand Movement) as to Figure (11-1) titled "VIC Switchover Circuit"

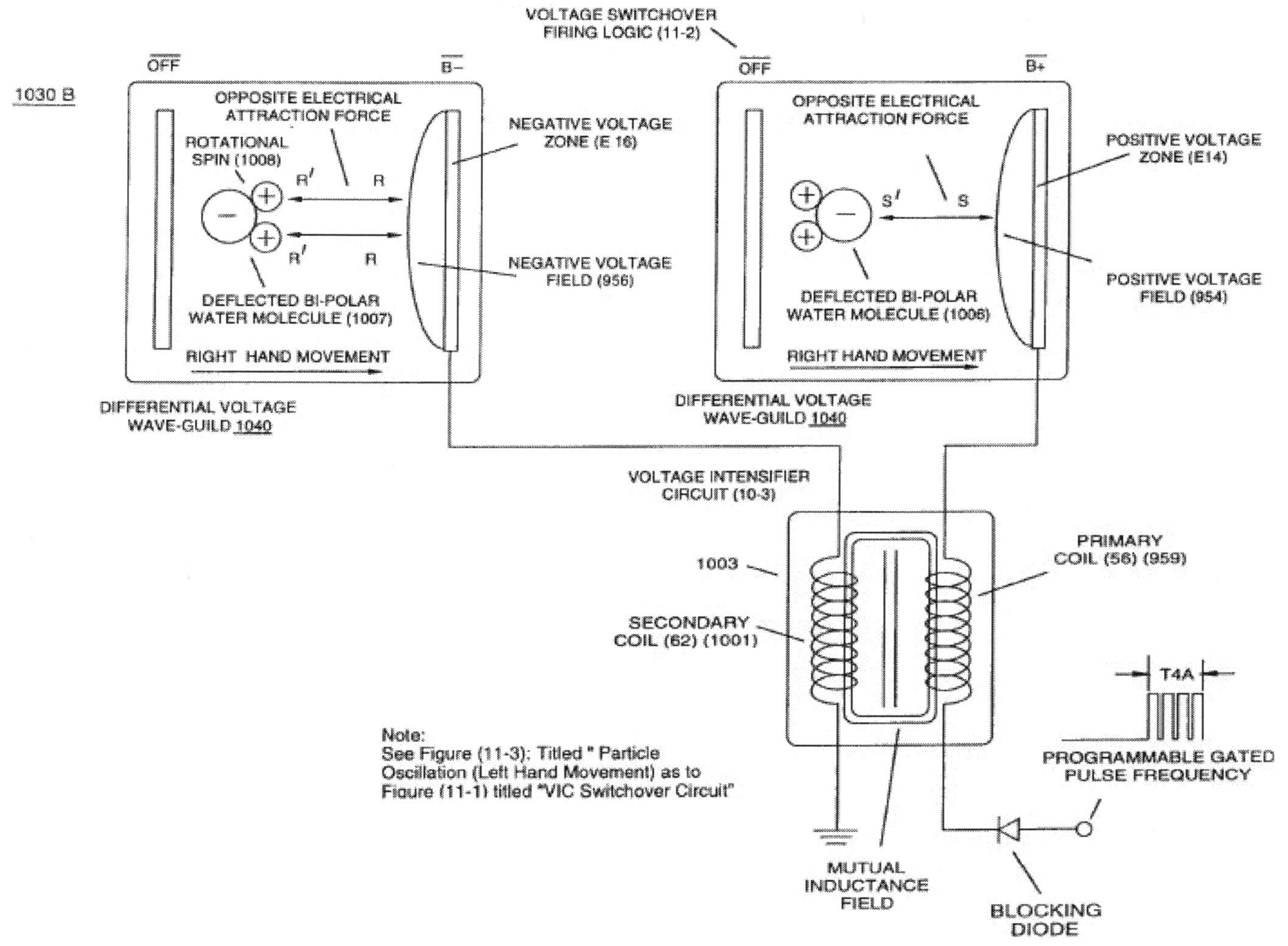

FIGURE 11-4: PARTICLE OSCILLATION (RIGHT HAND MOVEMENT)

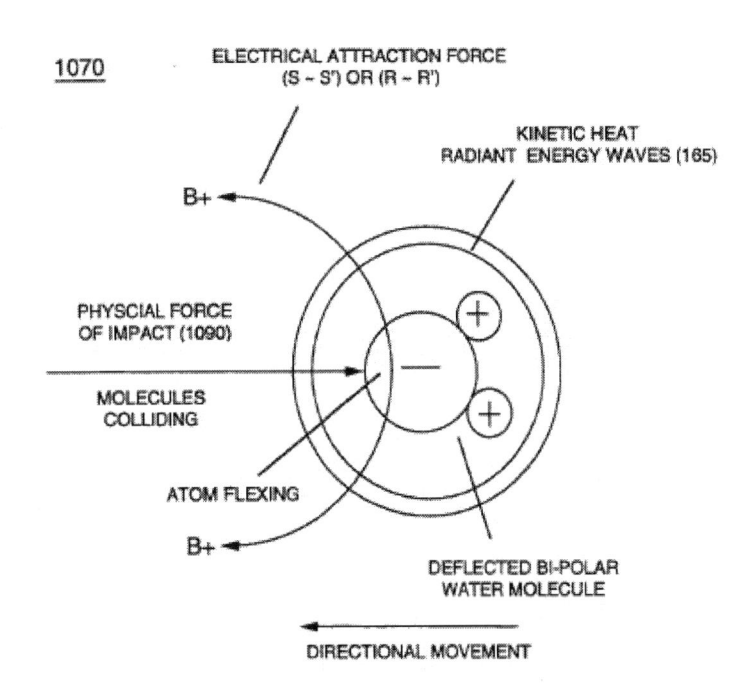

FIGURE 11-5: Particle Oscillation As An Energy Generator

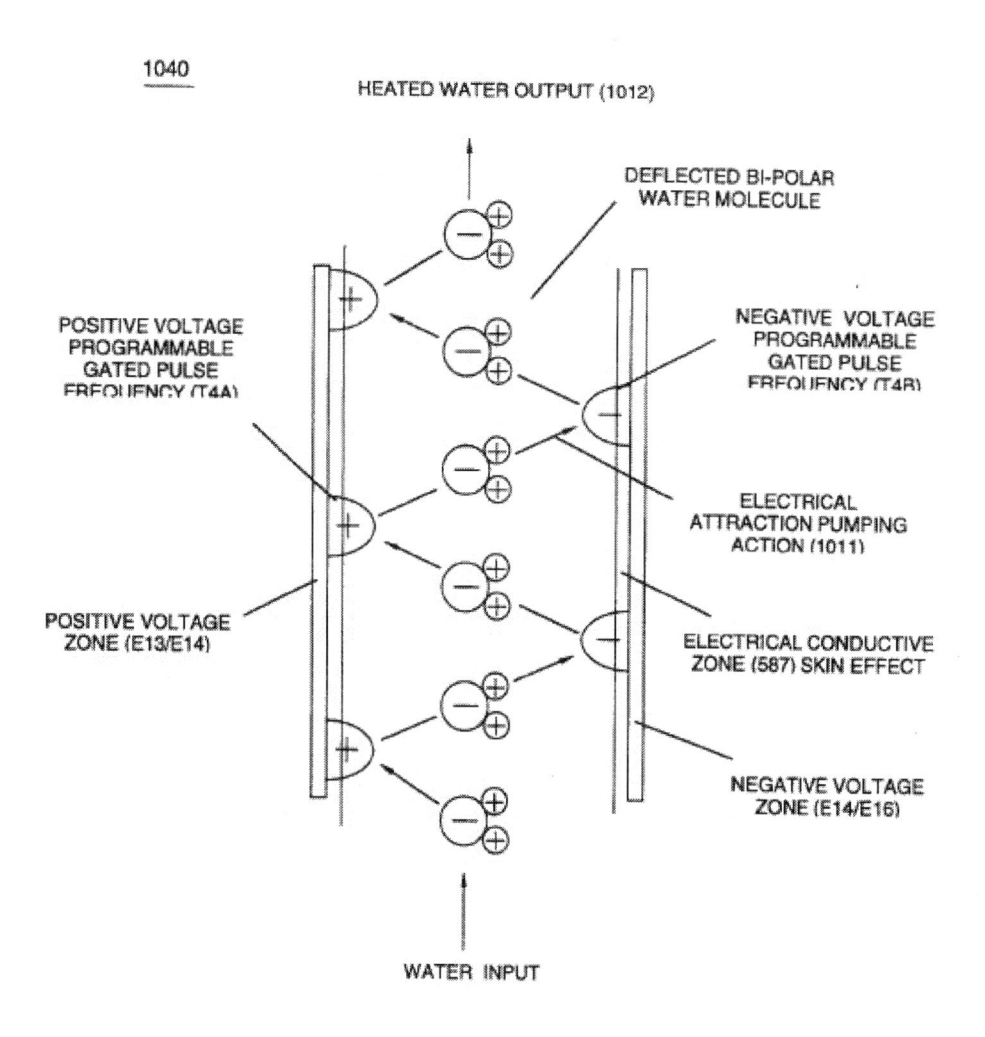

FIGURE 11-6: DIFFERENTIAL VOLTAGE WAVE-GUILD

RE: Steam Resonator   Memo WFC 430

1050

DIFFERENTIAL VOLTAGE WAVE-GUILD 1040A

DUAL VOLTAGE SWITCHOVER FIRING LOGIC (1013)

DIFFERENTIAL VOLTAGE WAVE-GUILD 1040B

1030 A $\frac{B-}{B+}$ OPPOSITE ELECTRICAL ATTRACTION FORCE $\frac{B-}{B+}$

$\frac{B+}{B-}$ OPPOSITE ELECTRICAL ATTRACTION FORCE $\frac{B-}{B+}$ 1030 B

ELECTRICAL REPELLING FORCE

ROTATIONAL SPIN (1019)

NEGATIVE VOLTAGE ZONE (E 16)

POSITIVE VOLTAGE ZONE (E18)

ELECTRICAL REPELLING FORCE

POSITIVE VOLTAGE ZONE (E14)

NEGATIVE VOLTAGE ZONE (E 17)

NEGATIVE VOLTAGE FIELD (1014)

DEFLECTING BI-POLAR WATER MOLECULE (1018)

NEGATIVE VOLTAGE FIELD (956)

POSITIVE VOLTAGE FIELD (1016)

DEFLECTING BI-POLAR WATER MOLECULE (1017)

POSITIVE VOLTAGE FIELD (954)

OSCILLATORY MOVEMENT

OSCILLATORY MOVEMENT

1015   1014

ELECTRICAL CROSSOVER SWITCH CIRCUIT (1060)

NOTE:
SEE FIGURE (3-46) TITLED "STEAM RESONATOR" AS TO FIGURE (32) TITLED "UTILIZING VOLTAGE POTENTIAL TO PRODUCE SUPERHEATED STEAM ON DEMAND"... WFC DEALERSHIP SALES MANUAL.

TRIGGER PULSE INPUT (1019)

1003

PRIMARY COIL (56) (959)

SECONDARY COIL (62) (1001)

T4A

PROGRAMMABLE GATED PULSE FREQUENCY

VOLTAGE INTENSIFIER CIRCUIT (10-3)

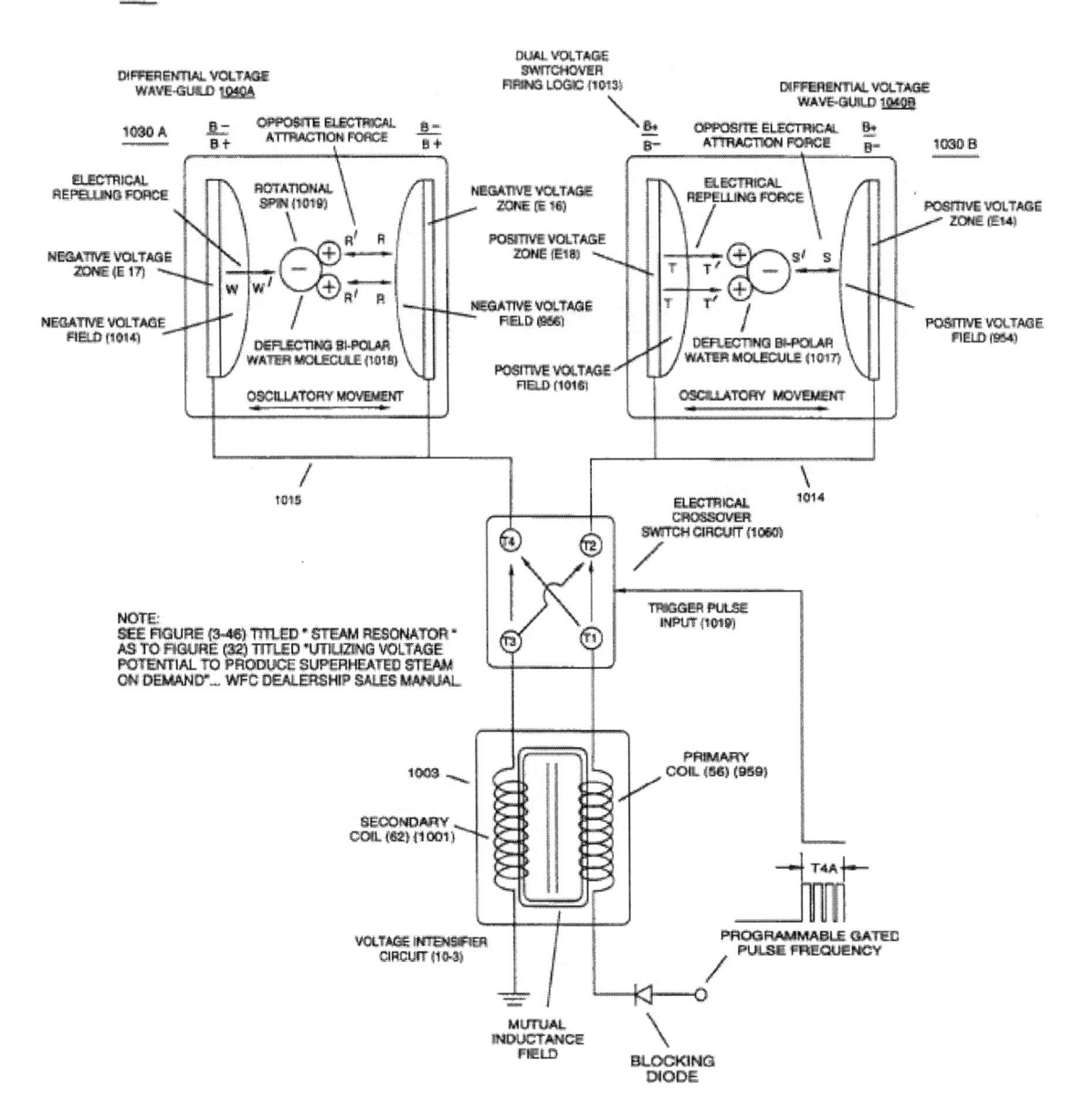

MUTUAL INDUCTANCE FIELD

BLOCKING DIODE

FIGURE 11-7: ELECTRICAL CROSSOVER SWITCHING CIRCUIT

Stanley A. Meyer

Appendix A

Table of Tabulations

Tab 33) Calculation on using "Mass Unit" to determine the amount of hydrogen contained in a gallon of Water.

Tab 34) Calculation on using "Mass Unit" to determined the amount of hydrogen contained in a gallon of Gasoline.

Tab 35) Calculation on using "Mass Unit" to determined the amount of hydrogen contained in a pound of Natural Gas vs. Water.

Tab 36) Calculation on using "Water as Fuel" to run a 50 hp I.C. Engine as compared to Gasoline.

Tab 37) Calculation on determining the liquid-volume of a "Water Droplet" per injection cycle.

Tab 38) Calculation on determining the electrical power input required to electrically energize the Voltage Intensifier Circuit per injection cycle.

Tab 39) Calculation on determining the liquid-volume of a "Water Droplet" required to run a 1000 Bhp I.C. Engine per injection cycle.

Section Appx A

RE: Table of Tabulation   Appendix A

Application Notes

Water vs. Fossil-Fuel Energy Content

Water is composed of (2) Hydrogen Atoms and (1) Oxygen Atom to form a molecule of Water.

(Tab 33)

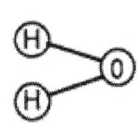

Atomic Mass Unit:
1 Electron (E) = 1 Proton (P) ~ 1Mu
Hydrogen Atom: 1E = 1P ~ 1Mu
Oxygen Atom: 8E = 8P ~ 8Mu
Atomic Mass Ratio (Mur) of Water
(2H X 1Mu) plus (1 Oxy. X 8 Mu) = 10 Mu's
** *See Appendix (B) Note (2)*

Whereby,
2H (Mu) divided by (10 Mu's) = 20%

Molecular Structure of Water
(Volumetric Displacement of Atom spheres)

Thus,
One gallon of Water contains 1.669 lbs. of Hydrogen

Energy-Yield Potential of Water

One water gallon equals 8.345 lbs

8.345 lbs x .20 = 1.669 pounds of Hydrogen / H20 gal.

1.669 pounds of hydrogen-fuel of water - .18359 lbs (11% per volume of impurities ...

typically 20 ppm ~ 40 ppm contaminates with Ambient Air being present) =

1.4854 lbs of hydrogen atoms available for gas combustion per gallon of Water approximately.

Water as Fuel ®

The by-product of burning gases derived from Water is environmentally safe since there is no carbon atoms present in the Water molecule ... resulting in the re-formation of Water "mist" after gas combustion ... being able to re-energize the newly formed Water Droplets for energy "reuse" once exposed to Sunlight. *(See Energy recycling graph 530 of Figure 5-6, once again)*

Stanley A. Meyer   Appx A 01

RE: Table of Tabulation   Appendix A

Gasoline is composed of (10) Carbon atoms and (8) Hydrogen atoms to form a gasoline molecule

(Tab 34)

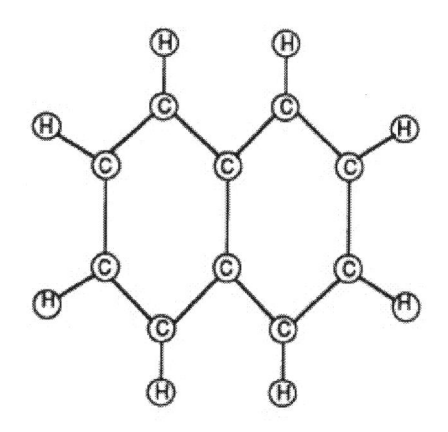

Molecular Structure of Gasoline
(Volumetric Displacement of Atom Spheres)

Atomic Mass Unit:
1 Electron = 1Proton ~ 1 Mu
Hydrogen Atom: 1E = 1P ~ 1 Mu
Carbon Atom: 6E = 6P ~ 6 Mu
Atomic Mass Ratio (Mur) of Gasoline:
(8 H X 1Mu) plus (10C X 6Mu) = 68 Mu's
** See Appendix (B) Note (2)

Whereby,
8H (Mu) divided by 68 (Mu's) = 11.7 %
Hydrogen Atoms

Thus,
One gallon of Gasoline equals 5.61 lbs/gal.
5.61 lbs/gal. times .117 = 0.656 lbs of
Hydrogen / Gasoline gal.

Fuel-contaminates: Distillation performance Point

Chromatogram of typical Gasoline:
degree C = (degree F - 32) / 1.8 @ 437 degrees F......10% / Volume impurities (Vi)

Therefore

.656 lbs of Hydrogen / Gasoline - .065 (Vi) = .591 lbs of Hydrogen Atoms available for Gas Combustion per gallon of Gasoline approximately.

Thermal Heat of Combustion

Water / gallon..........57,000 BTU'S approx.
Gasoline / gallon .....22,800 BTU'S approx.

Thereby

Water Energy-yield (Ey) is 2.5 times greater than Gasoline since the hydrogen content of water is more than twice that of fossil fuel of gasoline. (See U.S. National Bureau of Standards Monograph 168 (523 pages)(Feb.1981) Engineering Design Data Manual titled "Selected Properties of Hydrogen", CODEN NBSM A6 Library of Congress Catalog Card Number: 80-600195).

Stanley A. Meyer

RE: Table of Tabulation   Appendix A

Natural Gas is composed of (5) carbon atoms and (12) hydrogen Atoms to form a molecule of gas.

(Tab 35)

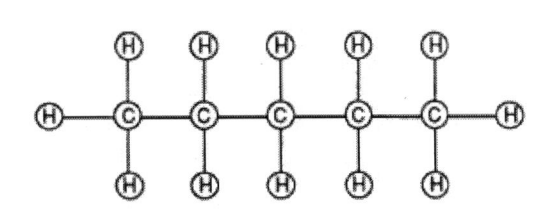

Atomic Mass Unit:
1 Electron (E) = 1 Proton (P) ~ 1 Mu
Hydrogen Atom: 1 E = 1P ~ 1Mu
Carbon Atom: 6 E = 6P ~ 6 Mu
Atomic Mass Ratio (Mur) of Natural Gas:
(12H x 1 Mu) plus (5C x 6Mu) = 42Mu's
** *See Appendix (B) Note (2)*

Molecular Structure Of Natural Gas
(Volumetric Displacement of Atom spheres)

Whereby
12H (Mu) divided by 42 (Mu's) = 28% of gas pound (lb).

Thus,
One pound (lb) of Natural Gas contains .28 lb of Hydrogen Atoms

Fuel Gas Contaminates: Cryogenic Processing:

12% Non-burnable Contaminates (carbon dioxide, heavy hydrocarbons, and Water vapor)

.28 lbs of hydrogen atoms x 12% = .28 lbs - .033 = .247 lbs Hydrogen atoms

Energy-Yield Potential:

.247 lbs hydrogen atoms - 10% (absorption Contaminates) = .247 - .024 = .223 lbs of hydrogen atoms available for gas combustion per pound of Natural Gas approximately.

Thereby

As to Normal Gas Burning Levels, One pound (1) lb of water contains approx. (.185) lbs of Hydrogen Atoms as compared to One pound (1) lb of Natural Gas which contains approx. (.223) lbs of Hydrogen Atoms. Water, of course, supplies its own oxygen to support the combustion process; whereas, Natural Gas must extract oxygen from air to produce thermal heat.

Energy Enhancement Process:

Energy Yield Enhancement of water is increase beyond Natural Gas burning rate by way of the Hydrogen Fracturing Process which simply prevents and/or retards the formation of the water molecule during thermal gas ignition/combustion... Energy priming the combustible gas atoms by stimulating the Atomic Energy Balance of Water (memo WFC 424) undergoing "Voltage Tickling of State Space" ...to cause "Particle Oscillation" as a "Energy Generator".

Stanley A. Meyer   Appx A 03

RE: Table of Tabulations   Appendix A

Gasoline vs. " Water as Fuel" : 50 hp Internal Combustion Engine   (Tab 36)

111 ml/ min. gasoline consumption rate (on-road tested) @ 65 mph ÷ 2.5 hydrogen-fuel

of water = 44.4 ml / min. water flow rate ÷ 60 sec. =

.740 ml /sec water-fuel consumption rate @ 65 m.p.h.

Water Injection Cycle   (Tab 37)

3,000 rpm ÷ 60 sec = 50 engine revolutions / sec ÷ 2 (Distributor Turn Ratio) =

25 Rotor revolutions / sec x 4 Water-Fuel Injectors = 100 Injection cycle / sec.

Therefore,

.740 ml / sec water-fuel rate ÷ 100 injection cycles / sec =

.0074 ml or 7.4 µl Water Droplet / injection cycle

Voltage Intensifier Circuit   (Tab 38)

40,000 volts @ 1 ma = 40 watts of applied electrical power

40 watts ÷ 12 volts battery = 3.3 amp/hr. (current) draw capacity

100 amp hr. battery ÷ 3.3 amp/hr. current consumption = 30.3 hr. battery-life

without recharging.

Mode of Operability   (Tab 39)

Example: 148 µl (1/8 Dia 2 cm length) Water Droplet ÷ 7.4 µl = 20 x 50 Bhp =

1000 Bhp I.C. Engine power-yield (gtnt) / injection cycle. (see Center for Electro-

magnetics Research, Northeastern University, Boston, MA. report titled "Powerful Water-

Plasma Explosion" as to Kansas State University report titled "Electrically Induced

Explosion in Water" affixed to WFC International Independent Test-Evaluation Report.

Remenber, water is 2.5 times more powerful (gtnt) than gasoline. (U.S. National Bureau of

Standards) ... as so established under U.S. Patent Security Laws 35 USC 101.

Stanley A. Meyer   Appx A 04

RE: Glossary of Application Notes   Appendix B

Note 1) The Electron Inhibiting Effect (631) of Figure (7-6) to cause "Electron Clustering" (Grouping/collecting negative charged particles at a given point) (700) of Figure (7-9) to produce "Negative Voltage Potential" (B-) at one side of Water Gap (Cp) of Figure (7-8) is accomplished by low electrical power input (Tab 38) when Choke-Coil (62) of Figure (7-1) magnetic field (FL2) (690) of Figure (7-8) during pulse on-time (49) impede "Electron-Flow" since electron mass is composed of electromagnetic matter which interacts with magnetic field strength (FL2). Capacitance Charging Effect (628) prevents amp influxing away from Water Gap (Cp) in a similar manner ... producing "Electrical Stress" (SS' ~ RR') (B+/B-) across Water Gap (Cp) since both Choke-Coils (56/62) conducts voltage potential (Negative or Positive) during pulsing operations.

Note 2) In determining volumetric sizing of the atom, Neutrons Clustering only enlarges the nucleus surface area since the additive Neutron (s) exhibits no electrical charge to deflect or change the orbital spin-velocity of the atom electrons.

Note 3) Universal Energy (9) of Figure (5-10) being a continuous energy potential (source) (C^2) coming into our space continuum and creating and sustaining/maintaining our expanding universe, as so extrapolated via mass equation $E=MC_2$. Whereby, Universal Energy ($C2$) having native intelligence to create mass (M) (to cause electromagnetic wave-vectoring ~ photon structuring ~ electron to proton grouping to form atoms ~ molecular arrangements to bring-on chemical processes to sustain life) which, in turns, emits radiant energy (E) under different stimuli conditions ... example, particle oscillation as a energy generator by way of "Electrical Stress".

Stanley A. Meyer   Appx B 01

Made in the USA
Columbia, SC
19 May 2022